Synthetic Sequences in Organic Chemistry

Drawing on the vast amount of experience the author has gained from working in industrial and university laboratories, this collection of excerpt reports contains essential details from literature relevant to the synthesis of compounds on a milligram to kilogram scale. The excerpts are composed using ChemDraw software and compiled in a Word document. A number of the compounds that have eluded efficient preparation in the past are now presented. Material which will improve any chemist's existing synthetic methodology can be found here. Each of the six chapters, with eighty excerpts, illustrates a novel application of syntheses selected from twenty-first-century literature.

Jerome Zoeller graduated from Rensselaer Polytechnic Institute, USA, with a B.S. in Chemistry and began work at the Rohm & Haas Co. in applications and process development with acrylic polymers. Graduate study followed at the University of Texas, USA, with research concentration in physical organic chemistry. Upon completing the Ph.D. program, he went on for two years as a Postdoctoral Research Fellow in the use of organometallics in synthetic and structural studies. At the Southwest Foundation for Research and Education he developed preparative methods for cortisol analogs, and their use in affinity chromatography and primate brain research. In this decade he lectured at Trinity University, USA. A change in career occurred in which he studied Fuel Science at The Pennsylvania State University, USA. This specialty was pursued at Texas A&M University, USA, in the use of specialized oxidative transformations in studies of fossil fuel structure. Upon retirement he moved to a startup pharma company, Targacept Inc., for synthetic work on nicotinic acetylcholine esterase ligands of potential neurological activity. Upon moving to Texas, he was prompted to compile some of the synthetic sequences displayed in the twenty-first century literature, for this book.

Synthetic Sequences in Organic Chemistry

Jerome Zoeller

CRC Press
Taylor & Francis Group
Boca Raton London New York

CRC Press is an imprint of the
Taylor & Francis Group, an **informa** business

First edition published 2023
by CRC Press
6000 Broken Sound Parkway NW, Suite 300, Boca Raton, FL 33487-2742

and by CRC Press
4 Park Square, Milton Park, Abingdon, Oxon, OX14 4RN

CRC Press is an imprint of Taylor & Francis Group, LLC

Library of Congress Cataloging-in-Publication Data
Names: Zoeller, Jerome, author.
Title: Synthetic sequences in graphical excerpts / Jerome Zoeller.
Description: First edition. | Boca Raton : CRC Press, 2023. | Includes
bibliographical references. | Summary: "Chemists faced with the need to
make a compound often have minimal exposure to the synthetic literature.
Even synthetic experts do not have every reaction at the tips of their
fingers. This unique volume contains an enormous range of synthetic
reactions from real examples"-- Provided by publisher.
Identifiers: LCCN 2022062031 (print) | LCCN 2022062032 (ebook) | ISBN
9781032502946 (hardback) | ISBN 9781032502939 (paperback) | ISBN
9781003397816 (ebook)
Subjects: LCSH: Organic compounds--Synthesis.
Classification: LCC QD262 .Z54 2023 (print) | LCC QD262 (ebook) | DDC
547/.2--dc23/eng20230512
LC record available at https://lccn.loc.gov/2022062031
LC ebook record available at https://lccn.loc.gov/2022062032

ISBN: 9781032502946 (hbk)
ISBN: 9781032502939 (pbk)
ISBN: 9781003397816 (ebk)

DOI: 10.1201/9781003397816

Typeset in Times
by Deanta Global Publishing Services, Chennai, India

Contents

Introduction

This is a collection of excerpts of journal articles, presented in sparse graphical form. The descriptions are from selected parts of the original reports, illustrating certain aspects of organic synthesis that are somewhere between conventional and ingenious.

Only superficial indications of reaction conditions are shown. The yields of the reactions are not usually indicated, because these depend on special techniques required, or because yields were based on recovered starting materials, or because they are frequently extrapolated from hplc analysis of the reaction mixture. Such caveats are added in the interest of not overpromising the yields that may be expected without reference to details in the original report.

The excerpts are collected in six chapters covering the types of target compounds, methods of synthesizing them, and variations on these methods.

Graphics pages are composed in ChemDraw, then imported into a WORD file.

1 Non-Natural Organic Compounds

The methods of natural product synthesis are used in the preparation of compounds for engineering or biomedical applications. Some of the applications include:

Fluorescent and isotopically labeled compounds for diagnostic imaging and tissue analysis.

A tripeptide incorporating an unnatural amino acid.

Special structures designed for electrooptical applications, nanomechanical scaffolds and switching devices, the chelation of ions, or for the catalysis of chemical transformations.

Organometallics with cytotoxic properties.

The modification of drugs for ease of synthesis, incorporation of bioisosteric functionalities, or addition of groups to improve the pharmacokinetic behavior of the drug.

DOI: 10.1201/9781003397816-1

THIENODIAZAPINONE ANTICANCER AGENTS

Lee, J.; Jung, H.; Kim, M.; Lee, E.;... Hah, J.-M. Discovery of novel 4-aryl-thieno[1,4[diazepin-2-one derivatives targetting multiple protein kinases as anticancer agents. *Bioorg.Med.Chem.* **2018**, *26*, 1628–1637

Several compounds in this structure class were very potent antiproliferative agents against two cell lines, with activity greater than Sorafenib.

One of th most promising kinase inhibitors prepared from **F** is **G**.

TETRAARYLMETHANE

Griffin, P.J.; Fara, M.A.; Whittaker, J.T.; Lolonko, K.J.; Catano, A.J. Synthesis of tetraarylmethanes via a Friedel–Crafts/desulfurization strategy. *Tetrahedron Lett.* **2018**, *59*, 3999

Tetraphenylmethane was to be liberated from diphenyldihydroanthracene **A** by removing linker **X**. The internal ring of **A** could be closed from tertiary alcohol **B**, functionalized with an **Aryl** group of choice, from the ketone precursor **C**. Such precursors are available from selected aldehydes or esters like **D**.

Retrosynthetic plan

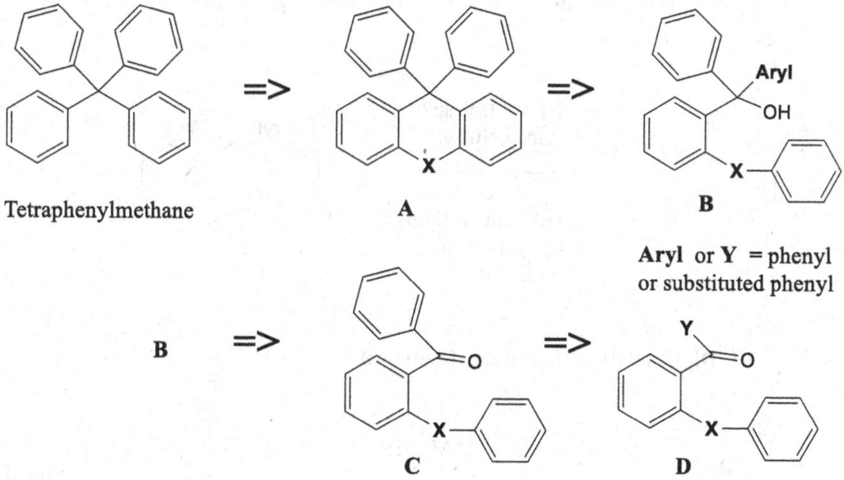

Tetraphenylmethane   **A**   **B**

Aryl or **Y** = phenyl or substituted phenyl

B => **C** => **D**

It had been found that stable intermediates such as 10,10 diaryl-5,10-dihydroanthracene type precursors could be prepared when two of the aryl groups were immobilized in an uncrowded coplanar orientation, as in **A**, with a short linking group **X**. In this report, **X** is sulfer, which can be removed from the structure, freeing a triarylmethane. Linking groups of other functionalities is conceptually promising in this application.

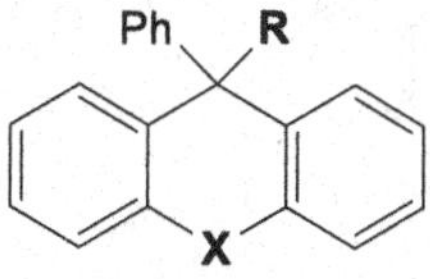

X	R	
CH$_2$	H	10-phenyl-dihydroanthracene
S	Aryl	precursor **A**

These transformations were realized in the two examples shown below in three steps.

The key precursor in both examples is prepared from phenylthiobenzene derivatives **D** and **H**. Both starting materials are arylated at the carbonyl group forming
the tertiary alcohols **E** and **I**. An intramolecular Friedel–Crafts reaction between
the phenylthio group and the tertiary alcohol then delivers the type **A** precursors **F**
and **J**. Desulfurization to the tetraarylmethanes **G** and **K** was problematic with both
nickel boride and Rainey nickel, requiring a tedious extractive workup.

Tetraphenylmethane itself

Diphenyl di-*p*-fluorophenylmethane

FLUORINATED LIQUID CRYSTAL

Kumon, T.; Hashishita, S.; Kida, T.: Yamada, S.; Ishihara, T.; Konno, T. Gram-scale preparation of negative-type liquid crystals with a –CF$_2$-CF$_2$- carbocycle unit via an improved short-step synthetic protocol. *Beilstein J. Org. Chem.* **2018**, *14*, 148–152

The liquid crystals

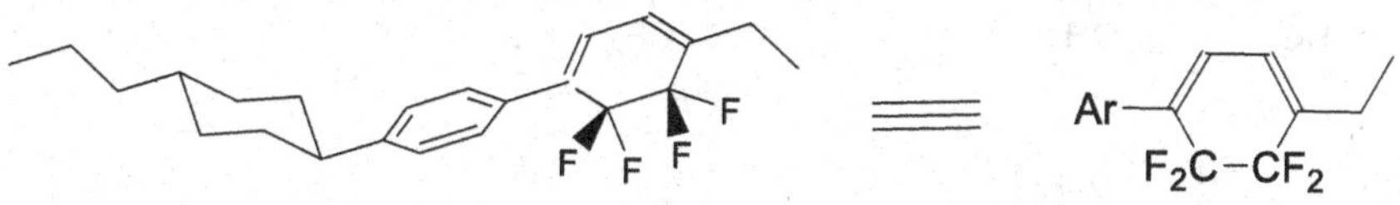

The materials here are electrooptical components with a large negative dielectric anisotropy (Δ ε) for vertical alignment display devices. The viscosity and stability of the mesophase are limitations on the properties of such liquid crystals. The practical use of these high perfomance compounds requires that they be available in industrial quantities. A scalable synthesis must therefore use high-yield, short-step synthetic protocols producing decagram quantities of an analytically pure product without the use of extreme temperatures, long reaction times, expensive starting materials, chromatographic purifications, or air-sensitive reagents.

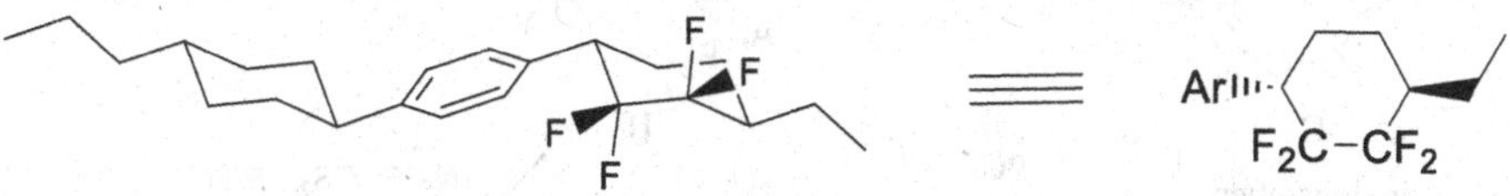

The use of organometallic reagents is a drawback of the synthesis shown here, yet these are two of the most promising candidates developed to date. Note the challenges in devising synthetic alternatives to i) the use of organometallics, ii) minimizing the formation of side product **F**, and iii) improving the final steps toward **B**.

$Ar = $

β-HYDROXYTYROSINE TRIPEPTIDE

Fernandez-Valparis, J.; Romea, P.; Urpi, F. Stereoselective synthesis of protected peptides containing an anti β-Hydroxy tyrosine. *Europ. J. Org. Chem.* **2019**, *2019*, 2745–2752

anti-β-hydroxy tyrosine   Protected tripeptide

Such peptides have been developed to realize biomedical objectives, such as proteolytic stability, entering the blood stream by passing through the intestine's brush border and the liver, inhibiting the angiogenesis of nascent tumors, acting as a selective protease, and crossing the blood brain barrier for neurotherapeutic effects.

In the appended words of the authors:

Direct, stereoselective, and catalyzed addition of a chiral N-azidoacetyl thiazolidinethione **C** to dialkyl acetals **D** followed by removal of the (chiral) scaffold with an amino ester (**E** to **G**), reduction of the azido group (**G** to **H**) and coupling with an amino acid (**I**) yields the tripeptide I containing the protected β-hydroxy tyrosine.

EXPERIMENTAL SEQUENCE

This tripeptide is assembled by first synthesizing the azido/allyl/benzyl/thiazolidi-nethione derivative **E** of the anti-β-hydroxy tyrosine.

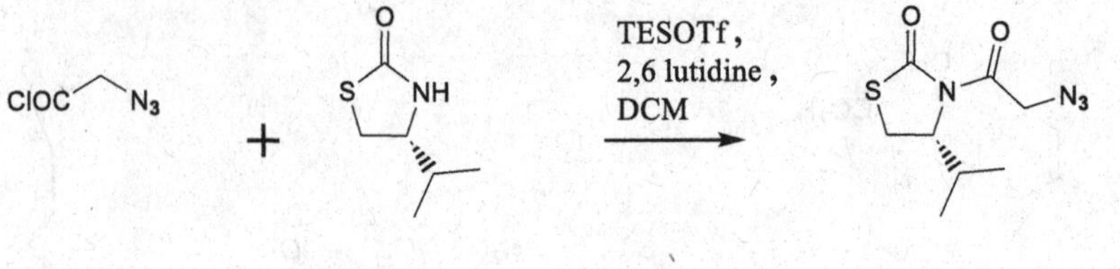

A   **B**   **C**
                                                              N-azidoacetyl thiazolidinone

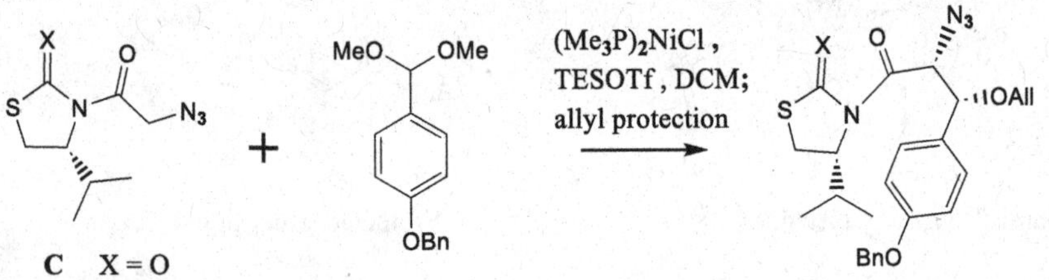

C X = O
      X = S
The N-azidoacetyl
thiazolidinethione
was used in some
experiments

D

E

86% dr = 95:5

Asparagine ester **F** is coupled with the carboxylate end of **E** by displacing the thiazolidinone in **E**, generating **G**.

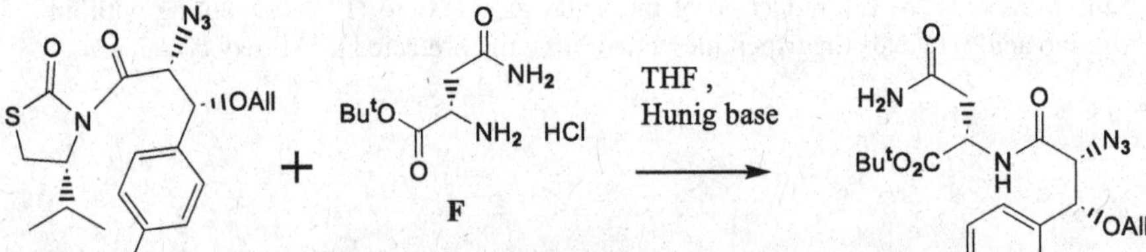

E

F

The chiral auxilliary in **E**
is displaced by the amine
in protected asparagine **F**.

G 60%

Upon reducing the azide group of **G** to the amine **H**, the doubly protected leucine derivative **I** is coupled with **H** forming the protected tripeptide.

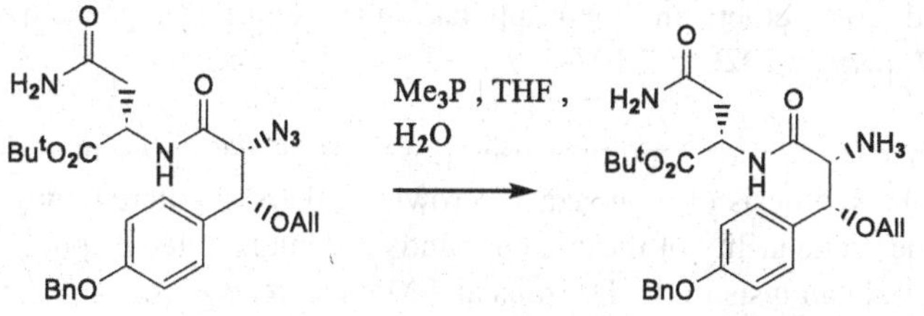

G

H 60%

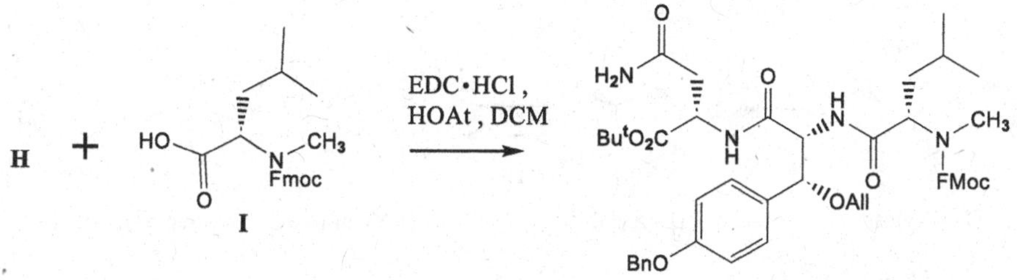

I

Protected tripeptide

FMOC	fluorenylmethyloxycarbonyl TESOTf triethylsilyl triflate
TESOTf	triethylsilyl triflate
DCM	dichloromethane Hunig base diisopropylethylamine
Hunig base	diisopropylethylamine
EDC•HCl	1-[3-(dimethylaminopropyl]-3ethylcarbodiimide hydrochloride
HOAt	I-hydroxy-7-azabenzotriazole

AZA CROWN ETHER: CHIRAL PTC AGENT

Nemcsok, T.; Rapi, Z.; Bagi, P.; Keglevich, G.; Bako, P. Synthesis of xylal- and arabinal-based crown ethers and their application as asymmetric phase transfer catalysts. *Chirality* **2020**, *32*, 107–119

Chiral crown ethers have found multiple uses in synthesis. The present study was directed toward the synthesis of monoaza 15-crown-5-ethers annulated with monosaccharide moieties. The utility of these compounds is demonstrated as phase transfer agents and chiral catalysts in the Darzans and Michael reactions.

SYNTHESIS OUTLINE

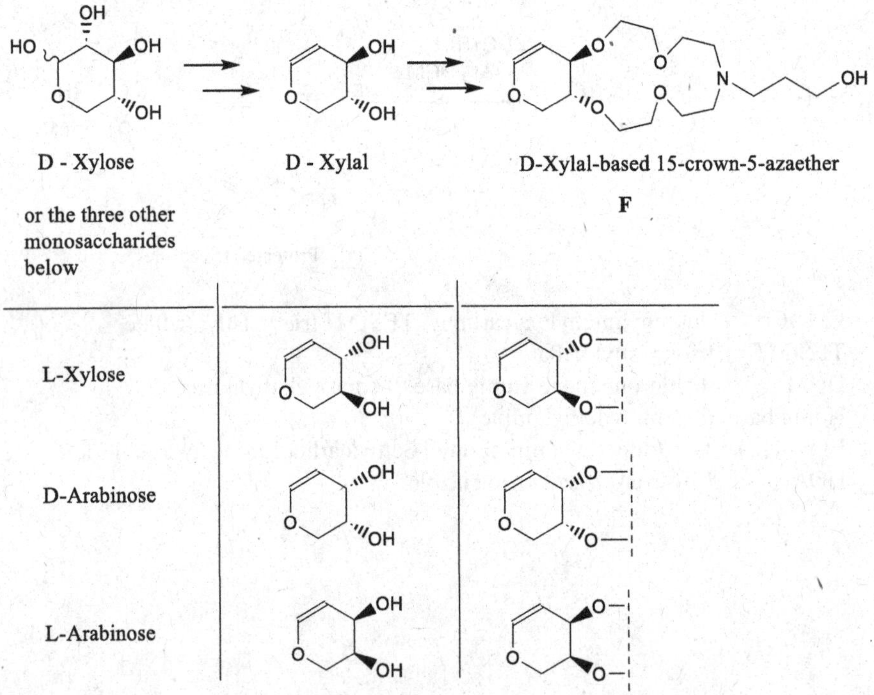

D - Xylose

or the three other monosaccharides below

D - Xylal

D-Xylal-based 15-crown-5-azaether

F

L-Xylose		
D-Arabinose		
L-Arabinose		

THE FOLLOWING SEQUENCE WAS USED FOR ALL FOUR MONOSACCHARIDES

Formation of the anomeric bromides **B**, and their elimination is a useful entry into the substituted pyrans, such as D-xylal and its monoaza crown ether **F**.

D - Xylose   **A**   **B**

Zn , CuSO4 , AcOH   5° C. 96%

NaOMe

C   **D - Xylal**

(Cl-CH$_2$CH$_2$-)$_2$O , 50% aq. NaOH , (Bu$_4$NH$^+$)$_2$SO$_4$$^{2-}$

Excess bis(2-chloroethyl) ether served as solvent in this PTC reaction.

H$_2$N-CH$_2$CH$_2$CH$_2$-OH , Na$_2$CO$_3$, MeCN

E   **F**

D-Xylal-based monoaza crown-ether. This diasteriomer was effective in the following two reactions.

EXAMPLES OF CATALYSIS WITH D-XYLAL BASED 15-CROWN-5-AZA ETHER

Benzaldehyde , 30% aq. NaOH , toluene , D-Xylal crown **F**

Darzans reaction

(2S,3R) epoxide 75 % yield , 85 % ee

Et$_2$O , THF , Na$_2$CO$_3$, H$_2$O D-Xylal crown **F**

Michael addition

(S) Michael adduct 56% yield , 34% ee

TRIPLE- ^{13}C LABELED PYRIDOXINE

"An example of a synthesis based on a mandatory starting material."

Bachmann, T.; Rychlik synthesis of ^{13}C$_3$-B6 vitamers labelled at three consecutive positions starting from ^{13}C$_3$-propionic acid. *Molecules* **2018**, *23*, #2117

Triple-isotope-labeled metabolites are used in forensic and diagnostic medicine to quantitate the levels of certain drugs in human subjects. The mass spectrometric method for this is based on spiking the analyte with a known amount of a triple-labeled standard. Comparison of the size of the parent peaks of labeled and unlabeled drugs gives the amount in the patient.

B6 vitamers triple-labeled with carbon-13 (^{13}C = •) were needed for metabolic studies. They were to be prepared from ^{13}C$_3$ labeled propionic acid **B**, which is one of the few triple-isotope-labeled starting materials available commercially. The challenges encountered in each step of the synthesis are described in detail, as well as the modified sequences and conditions that resulted in acceptable yields and purity of the intermediates and 13C$_3$ pyridoxine **PN**.

R^1	R^2	Vitamer	
-CH$_2$OH	-CH$_2$OH	**PN**	Pyridoxine
-CHO	-CH$_2$OH	**PL**	
-CH$_2$NH$_2$	-CH$_2$OH	**PM**	

B6 Vitamers

Triple ^{13}C labelled B6 Vitamer **PN**

^{13}C$_3$ Pyridoxine **PN**

The Synthesis :

SOCl$_2$;
Br$_2$ (1.5 equiv.) . ;
EtOH 0°

potassium
phthalimide
(2.5 equiv.) ;
MeCN , reflux

B
commercial

C

E

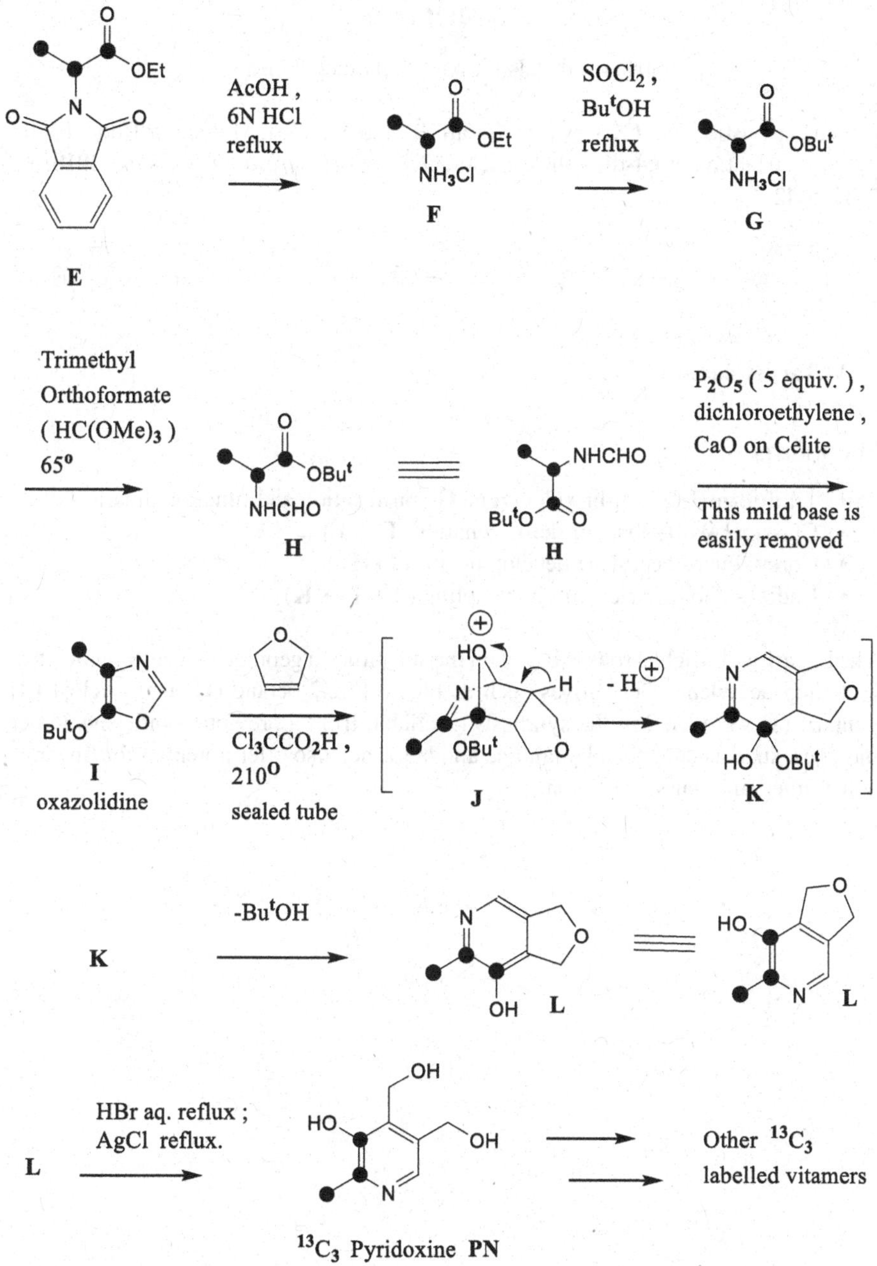

E

AcOH, 6N HCl reflux

F

SOCl$_2$, Bu^{t}OH reflux

G

Trimethyl Orthoformate (HC(OMe)$_3$) 65°

H ≡ H

P$_2$O$_5$ (5 equiv.), dichloroethylene, CaO on Celite

This mild base is easily removed

I
oxazolidine

Bu^{t}O

Cl$_3$CCO$_2$H, 210° sealed tube

J

− H$^+$

K

K

−Bu^{t}OH

L ≡ L

HBr aq. reflux ; AgCl reflux.

L

^{13}C$_3$ Pyridoxine PN

Other ^{13}C$_3$ labelled vitamers

NANOROD

"Structural rods for nanoscale mechanisms"

Kaleta, J.; Bastien, G.; Cisarova, P.; Batail, P., and Michl, J. Molecular Rods: Facile desymmetrization of 1,4-dimethylbicyclo[2.2.2]octane *Euro. J. Org. Chem.* **2018**, *37,* 5137–5142

M Target nanorod

Highlights

- 1,4 diformyl-[2.2.2]-bicyclooctane **G** construction and functionalization
- Classical Wolf–Kishner deoxygenation (**E** to **F**)
- Corey–Fuchs acetylene development (**G** to **H**)
- Cadiot–Chodkiewicz alkyne coupling (I + **J** → **K**)

The linearity of Michl's rods is based on the substitution geometry of certain moieties including acetylene. Derivatives such as bicyclo[2.2.2]octane (**H**) or bicyclo[1.1.1] pentane (**J**) were used in the synthesis (available from a previous synthesis). Other moieties, such as cubane, adamantane and benzene, also offer potential for linear or constrained angular substitution.

EtO₂C~~~CO₂Et → (MeO~~OMe, NaH) → **A** → (Br~~Br, NaH) → **B** → (HO~~OH, pTsOH, MeOH) → **C** → (LiAlH₄) → **D** → (1M HCl) → **E**

E → (H₂NNH₂, KOH) → **F** → (Swern [O]) → **G**

H 1,4-diethynylbicyclo[2.2.2]octane

(CBr₄ Ph₃P ; BuLi ; H₂O)

(BuLi ; TMSCl ; BuLi ; Br₂) → **I** / **J**

(BuLi; CuCl; then **I**) → **K** → (TBAF) → **L** → ([Pd], 4-bromopyridine) → **M**

BORAZAPHENANTHRENE

Abenzogar, A.; Sucunza, D.; Garcia-Garcia, P.; Vaquero, J. J. Remarkable effect of alkynyl substituents on the fluorescence properties of a B–N-phenanthrene. *Beilstein J Org. Chem.* **2019**, *15*, 1257–1261

Synthesis of five derivatives of 10a-bora-4a-azaphenanthrene. The polarity of the B–N bond imparts useful photophysical properties, which are enhanced by substitution with unsaturated or electron-donating functionalities.

R = –Cl   **B**

R =   **C**

R =   **D**

R =   **E**

R =   **F**

R = H   **A**   10a-bora-4a-azaphenanthrene

The syntheses

$+ \ H_2N$

[PdCl(allyl)]$_2$,
JohnPhos, Bu^{t}ONa,
Tol, 70°   82%

G

G →

BF_3K

SiCl$_4$, TEA, Tol
CPME, 110°
69%
Boron bridges the amine
with the styryl alkene

CPME = cyclopentyl methyl ether

H

Grubbs II cat. ,
DCM

I

B   R = Cl

The magnitude of the partial charges on the heteroatoms is influenced by the substituents R, which manifest themselves in the photophysical properties of the derivatives prepared below. Non-zero partial charges, in fact, occur at several of the carbon atoms involved in the conjugation.

Spectra of the UV-Vis absorption, and fluorescence emission of these derivative are shown, analyzed and interpreted.

This laboratory later reported the synthesis of another boraza-Polynuclear Aromatic **H**ydrocarbon.

Abenzogar, A.; Valencia, I.; Otarola, G.G.; Sucunza, D.; Garcia-Garcia, P.; Perez-Redondo, A.; Menduti, F.; Vaquero, J. J. Communication expanding the BN-embedded **PAH** family: 4a-aza-12a-borachrysene. *Chem. Comm.* **2020**, *56*, 3669–3672

12a-bora-4a-azachrysene   10a-bora-4a-azaphenanthrene   **A**

^{18}F-RADIOTRACER ROADMAP

Nieto, E.; Alajarin, Ramón; Álvarez-Builla, J.; Larrañaga, I.; Gorospe, E.; Pozo, M.A. A new and improved synthesis of the precursor of the hypoxia marker [^{18}F]-FMISO. *Synthesis* **2010**, *2010*, 3700

Four interlocking routes are explored for the large scale preparation of an imaging agent: the 18F radiotracer Fluoromoisonidazole **[18F]-"FMISO."**

Four Synthetic Routes

Starting Material **C**   Product **A**

2-nitroimidazole   **[^{18}F]-FMISO**

A diagram is presented to show a few synthetic routes for the preparation of a target compound. The routes are interconnected, much like a roadmap. The intermediates can be accessed by more than one route and can proceed toward an objective by alternate sequences.

The lack of stereocontrol, and the expense of starting material **C**, are aspects of all these routes that restrict the production and widespread diagnostic use of this promising imaging agent.

* This one pot acetonide opening and tosylation with tosyl anhydride and PPTS (**E** to **G**) is more efficient than the two-step route through diol **F**.

PPTS = Pyridinium p-toluene sulfonate

TECHNECIUM RADIOTRACER

A-85380 A drug with usage in analgesia and pain relief

Mori, D.; Kimura, H.; Kawashima, H.; Yagi, Y. … Saji, H. Development of ^{99m}Tc radiolabelled A85380 derivatives targeting cerebral nicotinic acetylcholine receptor: Novel radiopharmaceutical ligand ^{99m}Tc-A-YN-IDA-C4. *Bioorg. Med. Chem.* **2019**, *27*, 4200–4210

Technetium ^{99m}Tc-labeled radiotracers are routinely used for metabolic and imaging studies. Three chelates for attaching ^{99m}Tc to A-85380 were synthesized. The chelates consist of A-85380 linked to i) two acetates esters (**H, "IDA"**), ii) two pyridines (**G, "DDA"**), or iii) one of each (**F, "DAMA"**). A linker was designed to attach the drug to the chelates.

Such a tracer was needed in preclinical work on the experimental nicotinic acetylcholine esterase receptor-binding drug. Using different combinations of linking chains and chelates, several variants of the needed drug–chelate hybrids were prepared as follows.

PREPARATION OF THE CHLOROMETHYL DERIVATIVE OF A-85380

DIRECT ATTACHMENT OF THE THREE CHELATES TO THE LINKER-FREE CORE OF E

Y = CH$_2$Cl **E**

Y = Br **I** used in the preparation of longer linking chains

E + MeO$_2$C—N(H)—3-pyridyl $\xrightarrow{\text{K}_2\text{CO}_3,\ \text{KI}}$ Y = **F**

DAMA 11%

E + 3-Pyridyl—N(H)—3-pyridyl $\xrightarrow{\text{K}_2\text{CO}_3,\ \text{KI}}$ Y = **G**

DDA 54%

E + MeO$_2$C—N(H)—CO$_2$Me $\xrightarrow{\text{K}_2\text{CO}_3,\ \text{KI}}$ Y = **H**

IDA 60%

COORDINATION WITH RHENIUM TRICARBONYL TO EVALUATE PROCEDURAL CONDITIONS AND STABILITY

Rhenium is similar to technetium in its coordination behavior and stability constants with such ligands.

F $\xrightarrow{\text{10\% HCl, MeOH;}\ \text{1M NaOH, [Re(CO)}_3\text{(H}_2\text{O)}_3]\ \text{Br}\ 110°}$ Y = **Re-A-DAMA**

G $\xrightarrow{\text{[Re(CO)}_3\text{(H}_2\text{O)}_3]\ \text{Br}\ 110°}$ Y = **Re-A-DDA**

H $\xrightarrow{\text{10\% HCl, MeOH, 60°;}\ \text{1M NaOH, [Re(CO)}_3\text{(H}_2\text{O)}_3]\ \text{Br, 110°}}$ Y = **Re-A-IDA**

ATTACHMENT OF IDA CHELATE WITH LINKING CHAIN AND ^{99M}TC COORDINATION

It was found from the rhenium studies that the **IDA** chelates were ideal in this application when a short chain linkage was imposed between the drug and the chelate. Optimal overall results were achieved using a four-carbon alkyne for the linkage, and the coordinating conditions used for **Re-A-IDA**. The linkage and coordination of the ^{99m}Tc proceeded as follows.

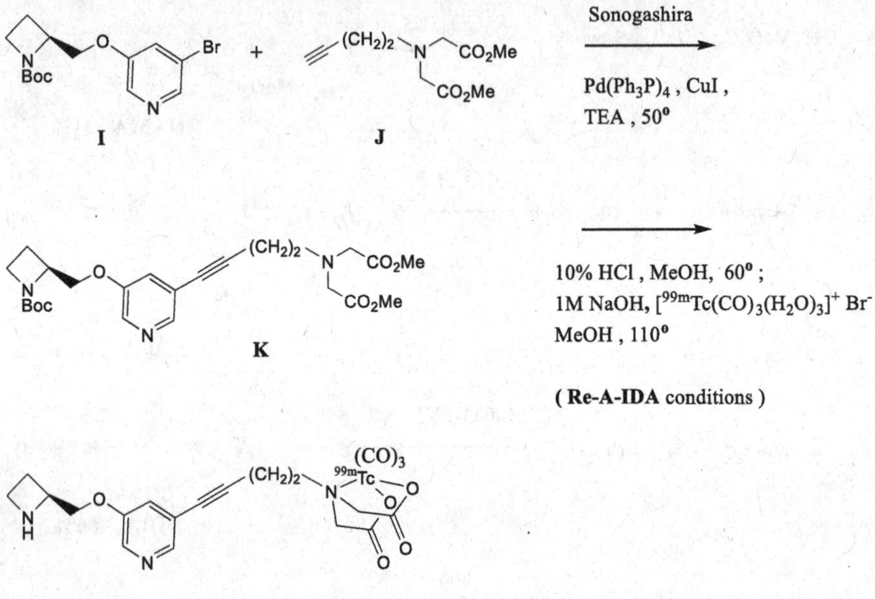

Sonogashira

$\longrightarrow$

Pd(Ph$_3$P)$_4$, CuI ,
TEA , 50°

I   **J**

$\longrightarrow$

10% HCl , MeOH, 60° ;
1M NaOH, [^{99m}Tc(CO)$_3$(H$_2$O)$_3$]$^+$ Br$^-$
MeOH , 110°

(**Re-A-IDA** conditions)

K

^{99m}**Tc-A-IDA**

CYTOTOXIC FERROCENE ANALOG

Agonigi, G.; Biancalana, L.; Lupo, M. G.; Montopoli, M.; Ferri, N.; Zacchini, S.; Binacci, F.; Biver, T.; Campanella, B.; Pampaloni, G.; Zamotti, V. and Marchetti, F. Exploring the anticancer potential of diiron bis-cyclopentadienyl complexes with bridging hydrocarbyl ligands: Behavior in aqueous media and *in vivo* cytotoxicity. *Organometallics* **2020**, *39*, 645-657

I IRON PENTACARBONYL DIMERIZATION

A simple example of the photochemical transformations of iron carbonyls.

II DISPLACEMENTS OF THE CARBONYLS IN THE DIIRON FERROCENE ANALOG A

One iron adds to the terminal carbon of the triple bond, which then closes onto the "free" carbonyl of the adjacent iron.

Acid catalyzed dehydration and rearrangement, with reconfiguration of the carbonyl.

III SUCCESSIVE REPLACEMENTS OF THE LIGANDS OF A

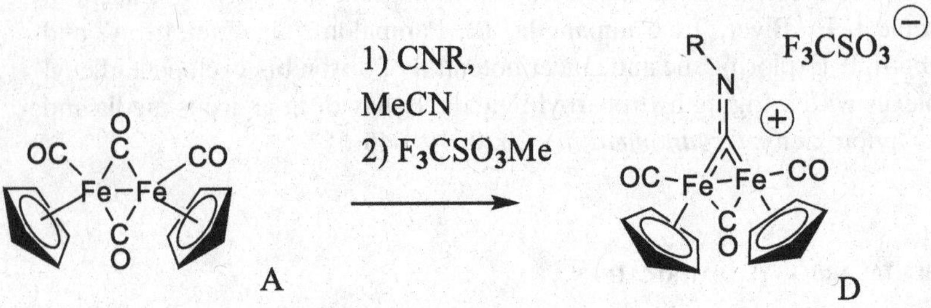

Isonitrile carbon displaces a "bridging" carbonyl.

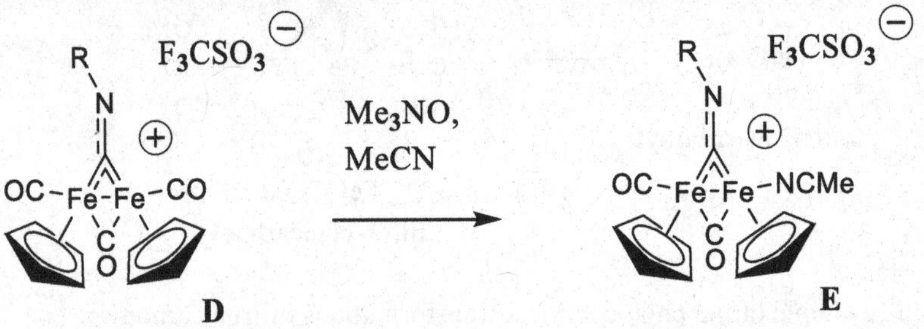

Ritter-type displacement of a "free" carbonyl ligand in **D** by an acetonitrile.

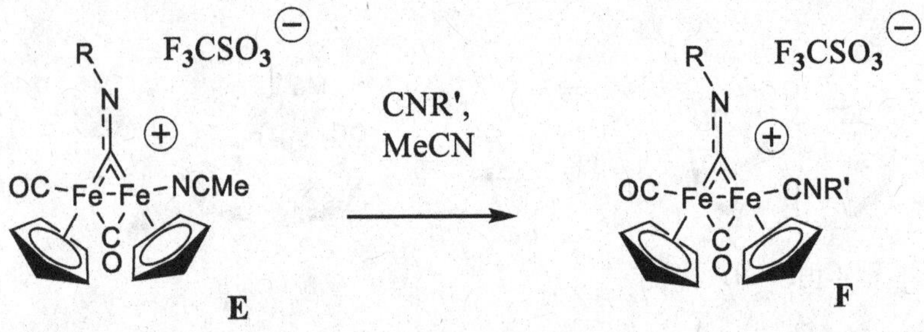

Displacement of the acetonitrile ligand in **E** by an isonitrile

Relevant physicochemical properties of a number of these diiron compounds were reported. These include D_2O solubility, log P_{ow} and IC_{50} cytotoxicity against three cancer cell lines; and iron uptake in the cell line MDA-MB-231 after incubation with diiron complexes. Compounds of type **D** and **F** were promising for *in vivo* cytotoxicity evaluation.

The cytotoxicity of such organometallics is thought to derive from the release of carbon monoxide at the tumor site, as reported in:

Carbon monoxide releasing molecules: The vasorelaxation effect. *Chemistry – A European Journal* **2022**, *28*, #20220228 and 2022016

EPATAZOCINE

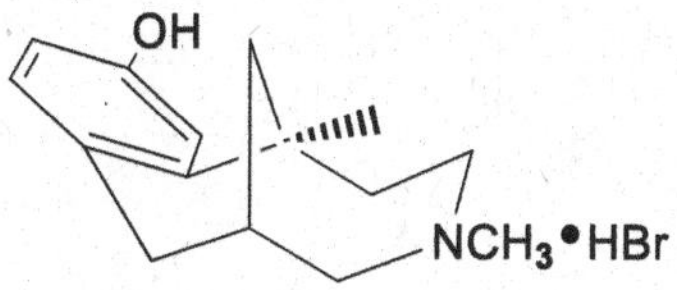

Li, R.; Liu, Z.; Chen, L.; Pan, J.; Lin, K.; Zhou, W. A novel and practical asymmetric synthesis of eptazocine hydrobromide. *Beilstein. J. of Org. Chem.* **2018**, *14*, 2340–2347

HIGHLIGHTS

- Improved commercial synthesis of eptazocine
- Asymmetric organocatalytic alkylation α to a carbonyl (**A** to **B**)
- Switching of the position of the carbonyl (**B** to **E**)
- Two carbonyl deoxygenations
- Troublesome benzylic-allylic methylene oxidation (**C** to **D**) and solvent problems with the Mannich cyclization (**E** to **F**)
- Apparent ambiguity in the graphical representation of chirality

RETROSYNTHETIC GUIDELINE

Retrosynthetic guideline

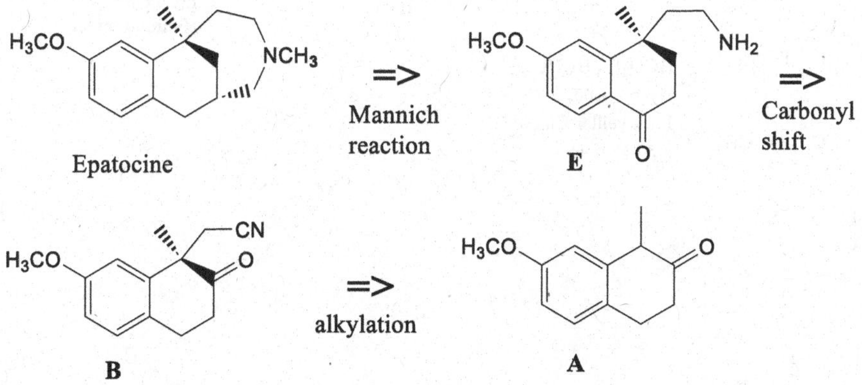

A seemingly circuitous, illogical and counterintuitive synthetic pathway, which was ultimately implemented as follows, is an adaptation to attempted procedures that went awry. Conditions for realizing the carbonyl shift (**B** to **E**) described in the retrosynthetic plan initially proved elusive.

F   P = H   HCO₂H, HCHO,
              H₂O   100%
G   P = CH₃   Eschweiller-Clarke
              methylation

PYRIDINE AZA CROWN
HYBRID CHELATE

"Prepared in the continuing search for selective ion chelates."

"PCTA [12]"

Enel, M.; Leygue, N.; Saffon, N.; Galaup, C.; Picard, C. Facile access to the 12-membered macrocyclic ligand PCTA and its derivatives with carboxylate, amiide, and phosphinate ligating functionalities. *Eur. J. Org. Chem.* **2018**, *2018*, 1765–1773

Specialty ligands are designed and screened for selectivity in binding with lanthanide, actinide and smaller ions for the purpose of isolation and purification.

In this sleek synthesis, the pyridine moiety can be substituted with functionalities with greater or lesser electron-donating effects for modulation of the stability constant of the ion complexes under study. Ligating functionalities other than the pendant carboxylic acids were also prepared.

2 Structural Moieties and Functional Groups

The transformation of functional groups is the basic bench-oriented task of organic synthesis. Each month, the literature describes not only new methods of realizing the classical transformations but also the preparation and chemistry of new functional groups and moieties. Excerpts in this chapter represent merely a sampling of this literature.

Every organic structure consists of a unique pattern of functional groups and moieties which express the physico-chemico-biomedical properties of the compound. Examples described in this chapter include:

The conversion of a thiocyanate into a diverse assortment of sulfur-containing functionalities.

Cyclization reactions involving the preparation of indenes, indolinones, carbolines, phthalic acid derivatives, naphthalimides, cylindracine C, a simple but synthetically elusive tetrahydroquinoline alkaloid, a pseudo aromatic pyran and a spiro-annulated butyrolactone.

The preparation and derivatization of undecylenic acid and phthalides.

Use of the pentafluorosulfanyl group as a bioisostere for the trifluoromethyl group to improve plasma stability of a drug.

The diastereoselective synthesis of cyclic 1,3,5 triols.

DOI: 10.1201/9781003397816-2

1,3-DIARYLINDENE SYNTHESIS

Kazakova, A.N.; Boyarskaya, I.A.; Panikorovskii, T.L.; Suslonov, V.V.; Khoroshilova, O.V.; Vasilyev, A.V. TfOH-Promoted Reaction of 2,4-Diaryl-1,1, 1-Trifluoromethyl-3-yn-2-oles with Arenes: Synthesis of 1,3-Diaryl-1-CF$_3$-Indenes and Versatility of Reaction Mechanisms. *Molecules* **2018**, *23*, 3079

1,3-diaryl-1-trifluoromethyl-indene

The preparation of a fluorinated propargyl alcohol **C** and its conversion into a fluorinated diarylindene product are described. The **C** to **D** conversion occurs through a series of carbonium ion rearrangements which culminate in a Friedel–Crafts attack on the benzene solvent and a final intramolecular arylation as outlined below.

A	F$_3$C-SiMe$_3$, CsF	With cautious use of CsF to avoid deprotection of TMS.
B		
	SnCl$_2$, MeCN or 6N HCl	Two alternative deprotection procedures were successful
C		
	TfOH, benzene	A plausible multistep cascade mechanism is proposed on the next page
D		

The annulation of **C** with benzene proceeds through the mesomeric cations **E** and **F**, which undergo nucleophilic attack by the added benzene, forming intermediate **G**. Rearrangement of **G** – with migration of highlighted hydrogen **H** – is followed by a second nucleophilic attack by the attached benzene ring on the attached benzene ring (**I** to **J**) which promptly eliminates the primed hydrogen H⁺ producing **D**.

The progress of these rearrangements seems to be influenced by the trifluoromethyl group.

D   1,3-diaryl-1-trifluoromethyl-indene

NITRILES FROM TERMINAL ALKYNES

Kori, R.; Murakami, K.; Nishiyama, Y.; Toma, T.; Satos. Copper-mediated conversion of alkynes into nitriles via iodotriazoles. *Chem. Pharm. Bull.* **2021**, *69*, 278–280

An attempt to generate an alkynyl azide **C** unexpectedly formed iodotriazole **D**. An azidization reaction on **D** yielded nitrile **F**.

The intermediacy of alkynyl azide **C** and azidotriazole **E** (shown below) is more speculative than plausible.

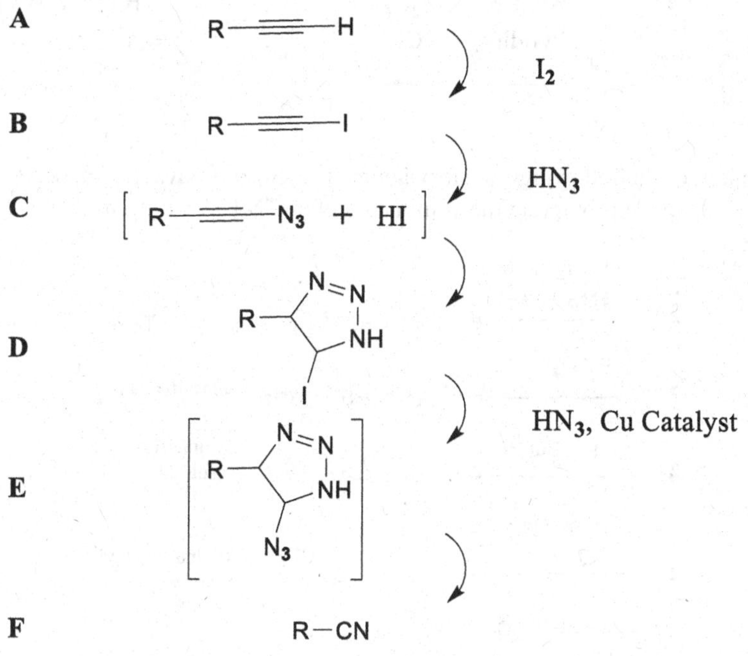

The realization of a palladium catalyzed arylation of **D** at the iodine would be useful in the synthesis of conformationally locked analogs, such as one of Combretastatin A-4, shown as **G** below.

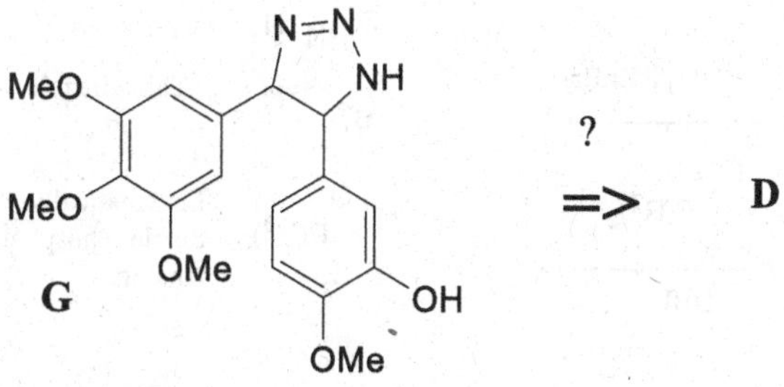

Combretastatin A-4 analog

THIOCYANATE TRANSFORMATIONS

Kong, D.-L.; Du, J.-X.; Chu, W.-M.; Ma, C.-Y.; Tao, J.-Y.; Feng, W.-H. Ag/ Pyridine Co-Mediated Oxidative Arylthiocyanation of Activated Alkenes. *Molecules* **2018**, *23*, 2727

N-phenylmethacrylamide **1** undergoes a Michael attack by thiocyanate ion forming the thiocyanate **2**, which can be converted into a variety of sulfur-based functionalities.

In many of the following transformations, the -SCN behaves as an electrophile, with a nucleophile prompting the elimination of a -CN leaving group.

ADDENDUM: The stable and inexpensive Langlois reagent as applied to this same starting material suggests a certain generality to this reaction.

Yang, F.; Klumphu, P.; Liang, Y.-M.; Lipshutz Copper-catalysed trifluoromethylation of N-arylacrylamides "on water" at room temperature. *Chem. Comm.* **2014**, *50*, 936–938

As with the thiocyanate, N-phenylmethacrylamide starting material 1 was used to prepare an analogous indolene-2-one derivative with a trifluoromethyl group instead of the thiocyanate functionality. A two-reagent combination – the Langlois reagent – was employed to realize the same free radical Michael addition of the trifluoromethylide radical to the alkene, with the same subsequent ring closure.

DEHYDROGENATIVE CROSS COUPLING

Shinde, V.N.; Dhiman, S.; Krishnan, R.; Kumar, D.; Kumar, A. Paper Synthesis of imidazopyridine-fused indoles via one-pot sequential Knoevenagle condensation and cross dehydrogenative coupling. *Org. Biomol. Chem.* **2016**, *16*, 6123–6132

Dienamines (**C** and **F**) are prepared from indole aldehydes (**A** and **E**). The highlighted bonds to hydrogen are then coupled using a bimetallic catalyst system. Tetracyclic carboline derivatives (**D** and **G**) are formed.

Piperidine , Toluene reflux
Claisen-Schmidt reaction

Pd(OAc)$_2$,
Cu(OAc)$_2$,
AcOH
Dehydrogenative
cyclization

Imidazo β-Carboline   Imidazo γ-Carboline

    The purview of the reaction of the hydrogens involved in the couplings at -**H** depends on their electrochemical environment, their geometrical relationship with each other, their proximity to the nitrogen atoms and the dual metallic catalysts. Electronic overlap, resonance effects and spatial proximity at the carbons bearing the two **H**s may be involved.

4-METHOXYPHTHALALDEHYDE

Moitessier, C.; Rifai, A.; Danjou, P.; Mallard, I.; Cazier Dennin, F. Efficient synthesis of 4-substituted-ortho-phthalaldehyde analogs: Towards the emergence of new building blocks. *Beilstein Journal of Organic Chemistry* **2019**, *15*, 721–726

C   D   F
          Expected   Obtained

This is a report of two quite unexpected reaction products from two successive reactions in sequence. In the first (above), an intramolecular Diels–Alder reaction on the propargyl furylmethyl ether **C** was expected to yield **D**. Instead, the product obtained was the annulated furan **F**. The preparation of **C**, and its conversion to **F**, is described below.

A   +   B   NaH , DMF   C

KOBu^t,
Bu^{t}OH ;
 then
MeI , NaH,
dioxane
──────→
 or
Ac$_2$O, Pyr.

C   D   E

F
R = Me
 or Ac.

Oxidation of this product **F** under three different conditions shown in the table selectively produced one or more of three different products, including the desired phthalaldehyde-4-methyl ether **G**, the unsubstituted ortho-phthalaldehyde **H** and hydroxy phthalic acid **I**. Notice the remarkable selectivity in these three reagents as a function of the particular combinations of the oxidizing agent and the allylic oxygen protecting group.

R	[O] Oxidizing Agent	Phthalaldehyde 4- Methyl Ether	"OPA" O-Phthalaldehyde	Hydroxyphthalic Acid
		G	**H**	**I**
-Me	SeO$_2$	0 %	0 %	100 %
-Ac	DDQ	0 %	100 %	0 %
-Me	DDQ	97 %	3 %	0 %

NAPHTHALIMIDE

"Simple compounds with remarkable agonistic or antagonistic
effects on enzymes."

Jin, C.; Alenazy, R.; Wang, Y.;Mowla, R.; … Ma. S. Design, synthesis and
evaluation of 5-methoxy-2,3-naphthalimide derivative as AcrB inhibitors for
the reversal of bacterial resistance. *Bioorg. Med. Chem. Lett.* **2019**, *29*, 882-889

An unconventional ring annulation method is used to access a substituted naphthalimide which was found to exert synergistic effects with certain antibiotics.

The methylation and bromination of hydroxy xylene **A** under mild free radical conditions produce the tetrabromo product **B**. Condensation of **B** with maleic
anhydride under KI catalysis unexpectedly cascaded through successive Michael
additions and dehydrobrominations to form the methoxy benzo phthalic acid **C** in a
single step. Conversion into naphthalimide **E** followed.

Compound **E** renders certain infectious agents susceptible to cell death when
treated with antibiotics.

Addendum: A recent addition to this class of compounds has been reported.

LiZhang, P.; Gopalas, L.; LinZhang, S.; Caia, G.-X.; Zoua, C.H. An unanticipated
discovery towards novel naphthalimide corbelled aminothiazoximes as potential
anti-MRSA agents and allosteric modulators for PBP2a *Euro. J. Med. Chem.* **2022**,
229, #114050.

GLUTAMATE NEUROANTAGONIST

K Newly developed G-protein antagonist.

Brumfield, S.; Korakas, P.; Silverman, L.S.; Tulshian, D. … Li, CSynthesis and SAR development of novel mGluR1 antagonists for the treatment of chronic pain. *Bioorg. and Med. Chem. Lett.*, **2012**, 22, 7223–7226

HIGHLIGHTS

- Successive pyrimidine substitutions and refunctionalizations (**A** to **E**)
- Closing of a dihydrothiophene ring (**E** to **F**)
- Displacement of methylthiocarboxylate group by dimethylamine (**F** to **G**)
- Ambiguity in the mode of addition of p-methoxyaniline (**H** to **K**)

A

POCl₃ , DMF

double chlorination unavoidable

B

H₂NOH , AcOH ;
SOCl₂ reflux

intended nitrile formation does not proceed without the unwanted hydroxyl formation

C

POCl₃ , reflux

replacement of the lost chlorine

D

the two chlorines in **D** are then successively functionalized as follows

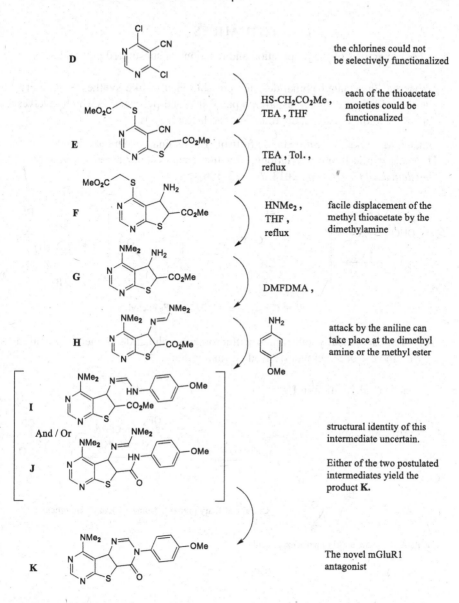

D

the chlorines could not
be selectively functionalized

HS-CH₂CO₂Me ,
TEA , THF

each of the thioacetate
moieties could be
functionalized

E

TEA , Tol. ,
reflux

F

HNMe₂ ,
THF ,
reflux

facile displacement of the
methyl thioacetate by the
dimethylamine

G

DMFDMA ,

H

attack by the aniline can
take place at the dimethyl
amine or the methyl ester

I

And / Or

J

structural identity of this
intermediate uncertain.

Either of the two postulated
intermediates yield the
product **K.**

K

The novel mGluR1
antagonist

PHTHALIDES

"Four examples of the preparation and reactions of substituted phthalides."

I Racemic fluorinated phthalides, for possible Hauser-like syntheses of fluorinated medicinal compounds, are prepared in a one-pot reaction which involves addition to the aldehyde, deprotection and lactonization.

Innaba, M.; Sakai, T.; Shinada, S.; Sulishi. T.; Nishina, Y.; Shibata, N.; Amii, H. Nucleophilic fluoro alkylation/cyclization route to fluorinated phthalides. *Beilstein J. of Org. Chem.* **2018**, *14*, 182–186

$$R^F = (CF_2)_nF \ (\ n = 1, 2, 3\ ) \ or \ F_5 \ Phenyl$$

II A stereospecifically substituted trifluoromethyl phthalide isomer is prepared with the use of a menthol-derived chiral auxiliary.

Innaba et al. citation in **I**

Chiral auxilliary prepared from (+)-benzylaminomenthol

mCPBA =   meta-chloroperbenzoic acid

III Unsubstituted phthalide is converted to the homoenolate with a unique pair of bases. A chiral catalyst guides its stereoselective Michael addition to an α,β, unsaturated carbonyl group, forming **G**.

Sicignano, M.; Schettini, R.; Sica, L.; Pierri, G.; De Riccardis, F.; Izzo, I.; Maity, B.; Minenkov, Y.; Cavallo, L.; Della Sala, G. Unprecedented diastereoselective arylogous Michael addition of unactivated phthalides. *Chem. Europ. J.* **2019**, *25*, 7131–7141

IV Base catalyzed reactions on chiral substituted lactones like **H** produce Hauser-like products such as **J** with retention of configuration.

Chakraborty, S.; Das, G.; Ghosh, S.; Mal, D. Regioselective syntheis of napht hoquinone/naphthoquinol-carbohydrate hybrids using the Hauser annulation as the key synthetic step has been described. *Org. Biomol. Chem.* **2016**, *14*, 10636–10647

H   **J**

10-UNDECYLENIC ACID

"The preparation and functionalization of 10-undecylenic acid."

HIGHLIGHTS

- Undecylenic acid from castor oil by pyrolysis
- A useful long chain synthon or linker **A**, functionalized at α and ω
- Functionalization of the alkene group
- Cryptochiral compounds

10-UNDECYLENIC ACID FROM CASTOR OIL

Chem. Berichte **1877**, *10*, 2035

Perkins, G. A.; Cruz, A. 0. Synthesis of Compounds Similar to Chaulmoogric Acid. II dl-Chalmoogric Acid. *J Am. Chem. Soc.* **1927**, *49*, 1070–1077

The crude undecylenic acid is obtained by pyrolysis of castor oil, followed by ozonation to the aldehyde and bisulfite purification as below.

OXIDATION AND PURIFICATION OF THE ALDEHYDE A

Brunner, A.; Hintermann, L. Configurational Assignment of "Cryptochiral" 10-Hydroxystearic Acid Through an Asymmetric Catalytic Synthesis. *Helv. Chem. Acta* **2016**, *99*, 928–943

O_3, DCM
-78°;
Me_2S

HO_2C—$(CH_2)_7$— (Undecylenic Acid)

HO_2C—$(CH_2)_7$—R crude impure C_{10} acid aldehyde

R = ⸾⸾⸾CHO   **A**

R = ⸾⸾⸾ $\overset{OH}{\underset{SO_3Na}{}}$ **B**

Purify through $NaHSO_3$ adduct in buffer

HO_2C—$(CH_2)_7$—CHO   **A**
C_{10} acid ω aldehyde

Regenerate aldehyde with CH_2O (7 equiv.), Extract into TBME

TBME = *tert*-butyl methyl ether
DCM = dichloromethane

FORMATION OF A CRYPTOCHIRAL ALCOHOL FROM THE ALDEHYDE

D   Cryptochiral Secondary Alcohol,
10-(S)-Hydroxy stearic Acid

THE TERMINAL ALKYNE FROM UNDECYCLENIC ACID

C_{11} acid adehyde **G**

ISOPHOS = 1-isopropyl-(2S,5S)-2,5-dimethylpholane

METATHESIS FORMING AN ω-HYDROXYL CRYPTOCHIRAL C_{15} CARBOXYLIC ACID

Muñoz, L.; Bosch, M.P.; Rosell, G.; Guerro, A. Asymmetric synthesis of (R) and (S)-4-methyloctanoic acids. A new route to chiral fatty acids with remote stereocenters. *Tet. Assym.* **2009**, *20*, 420–424

This cryptochiral **J** is formally 12-(S)-methyl-15-hydroxy-pentadecyl carboxylic acid.

PENTAFLUOROSULFANYL – SARM

Shao, P.; Zhou, Y.; Yang, D.; Wang, M.; Lu, W.; Jin, J. Synthesis of Aryl Propionamide Scaffold Containg a Pentafluorosulfanyl Moiety as SARMs. *Molecules (Basel, Switzerland)* **2019**, *24*, 4227

The pentafluorosulfanyl group -SF$_5$ is a bioisostere of -CF$_3$. The two are similar in size, electron withdrawing effect and biomedicinal activity, but the -SF$_5$ is comparatively resistant to metabolic degradation. Several derivatives of osterine A (a **S**elective **A**ndrogen **R**eceptor **M**odulator, **SARM**) with -SF$_5$ in lieu of CF$_3$, and different substituents on the phenyl rings, are targeted in this synthesis.

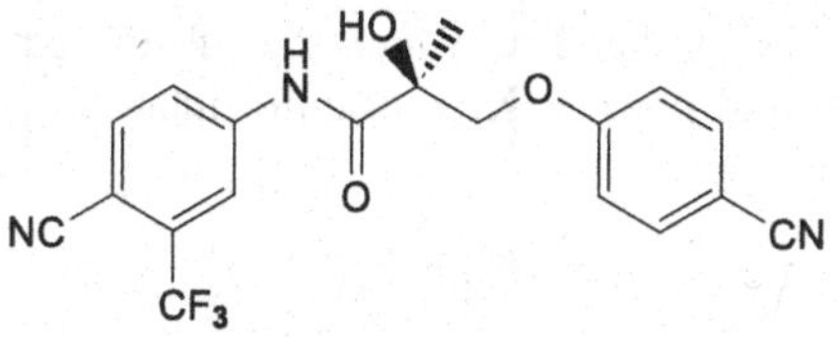

SARM Osterine A

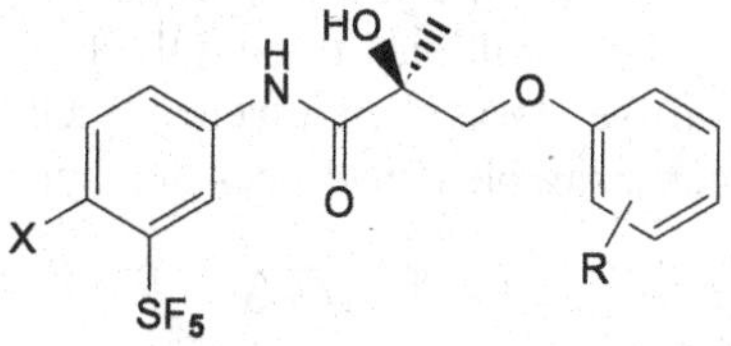

M Targeted Derivatives

X = CN
R = EWGs and EDGs

The pentafluorosulfanyl aniline moiety (**D**) is prepared as below from m-pentafluorosulfer nitrobenzene (**B**). The propionamide linkage is formed using (R)-proline (**F**) as a chiral auxilliary and methacryloyl chloride (**E**) as starting material, as shown below and on the next page.

The m-Pentafluorosulfanyl Aniline Group

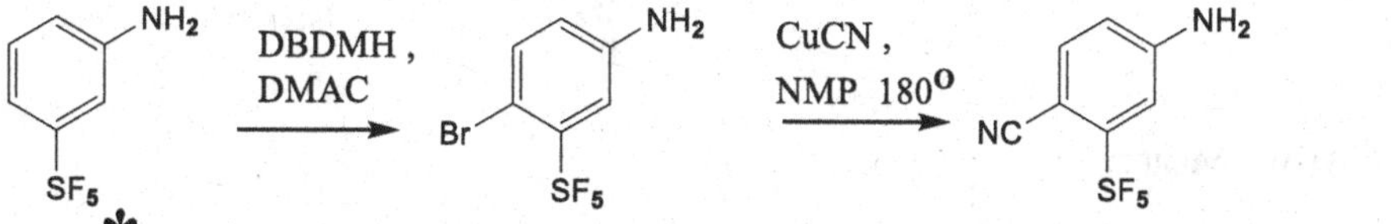

B *

C

D

used in reaction with **I**
in the following sequence

DBDMH = 1,3-dibromo-5,5-dimethylhydantoin
DMAC = N,N-dimethylacetamide
NMP   = N-methyl pyrrolidinone

* 3-nitrophenysulfur pentafluoride is commercially available

PROPIONAMIDE LINKAGE

The stereochemical transformation of **G** to **H** involves the intermediate bromonium ion **G** • The chirality of that resultant bromonium ion is influenced by the chirality of the carboxylic acid in **G**. Subsequent intramolecular S_N2 attack by the carboxylate oxygen manifests that chirality into **H**. The proline auxiliary is removed under harsh acidic conditions, freeing the bifunctional chiral linker **I**, to which the substituted aniline (**D**) is then attached, yielding the desired product **K**, and, unfortunately, a nonseparable dehydrogenated byproduct **J**.

ARYLOXY MOIETY

A series of phenoxides **M** are prepared, from which the undesired adducts can be separated.

L

The unexpected partial loss of -CN led to two derivatives of **M**. Only the derivative with X = CN has the requisite **SARM** activity as a muscle relaxant.

M Targeted Derivatives

X = CN , H   separable

R = EWG or EDG

SPIROLACTONE

Huo, J.Q.; Fan, Q.F.; Wu, X.; Li, X.; Zhou, S.; Xiong, L.X.; Kalinina, T.; Glukhareva, T. Efficient construction of bioactive trans-$5_A5_B6_c$ spirolactones via bicyclo[4.3.0] a-hydroxy ketones. *Org. Biomol. Chem.* **2018**, *16*, 1163–1166

This synthesis of the target spirolactone **M** is of possible utility in the synthesis of compounds like abissomycin and galiellalactone. The spiro-annulated lactone attached to a carbocycle at a quaternary chiral center is the challenging moiety of interest.

| Abyssomycin | Galeillalactone | **M** target spirolactone |

The indene precursor **D** was efficiently assembled in three steps from **A**. Manipulation of the oxygen functionalities then proceeded through two steps (**D** to **E/F**, then to **G/H**), in which the intermediate mixtures of products could not be separated. Intermediate **I** (separable) was ultimately obtained, with the key stereochemistry needed for installation of the angular spirolactone (**I** to **M**).

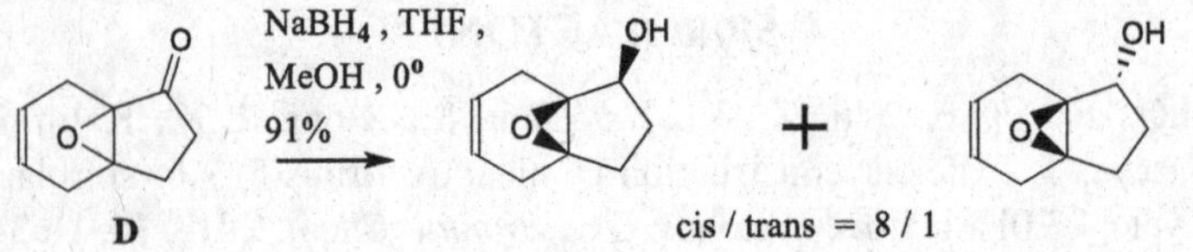

The epoxide was found to survive the borohydride attack, but a better enantioselectivity was needed. A new route through **E** and **F** led to inseparable pairs, which could be used to complete the synthesis.

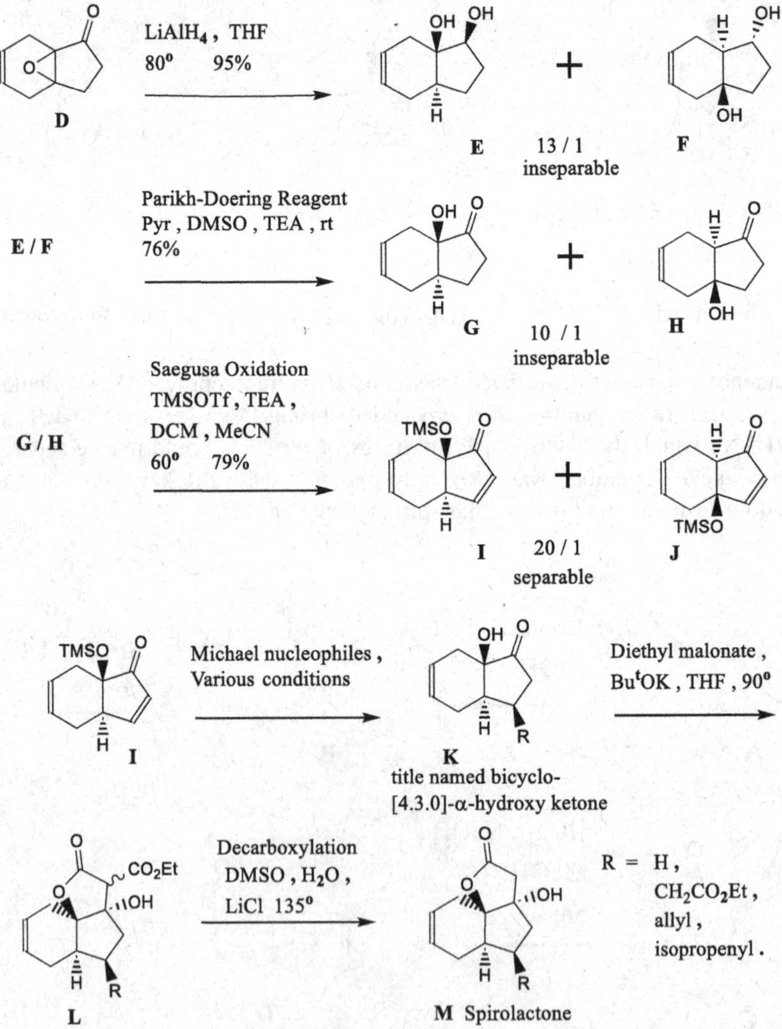

R = H,
CH$_2$CO$_2$Et,
allyl,
isopropenyl.

TETRAHYDROQUINOLINE (THQ) SYNTHESIS

Jagdale, A.R.; Reddy, R.S.; Sudalai, A. A concise enantioselective synthesis of 1-[3(S)-(dimethylamino)-3,4-dihydro-6,7-dimethoxyquinoline-l-(2H)-yl]-propan-1-one. *Tetrahedron Assym.* **2009**, *20* (3), 335–339

L Target THQ

The target THQ is the inotropic drug S-903, used to alleviate acute respiratory distress syndrome.

RETROSYNTHETIC PLAN

The 3(R)-hydroxy tetrahydroquinoline **G** would be inverted at the alcohol and acylated at the nitrogen to form the target **L**. Sulfite **C** incorporates the needed chirality for the generation of **G**, and **C** itself is available from a simple cinnamate **A** through sharpless asymmetric dihydroxylation, nitration and sulfite formation.

This conversion of **C** to **G** is characterized by the single step reduction of cyclic sulfite **C** across four functional group transformations illustrated on the following page.

Cascade conversion of the sulfite, C to G. Mechanistic continuity is not implied.

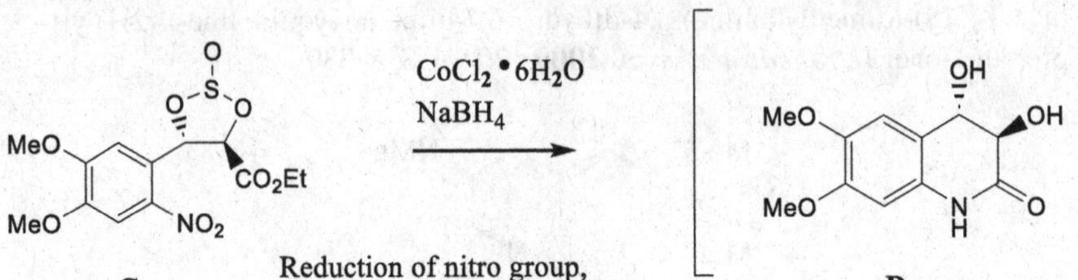

C
From cinnamate A

CoCl$_2$ • 6H$_2$O
NaBH$_4$

Reduction of nitro group, amide formation, and sulfite hydrolysis.

D

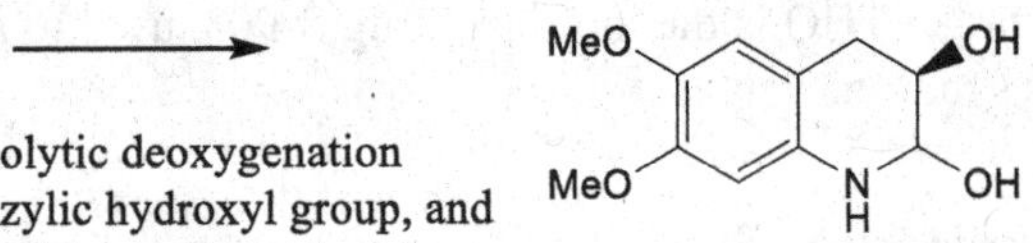

D → Hydrogenolytic deoxygenation of the benzylic hydroxyl group, and reduction of the amide carbonyl.

E

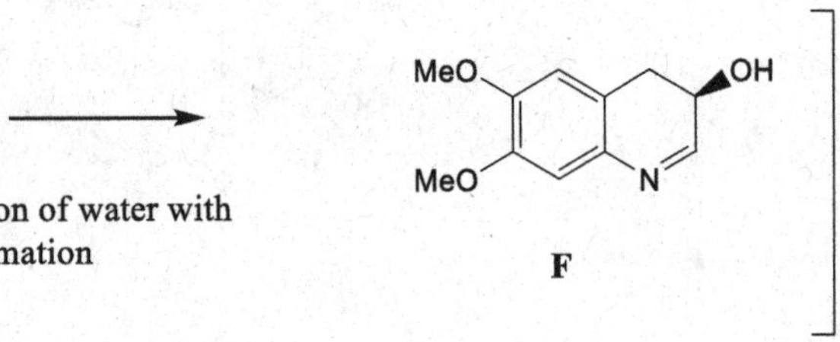

E → Elimination of water with imine formation

F

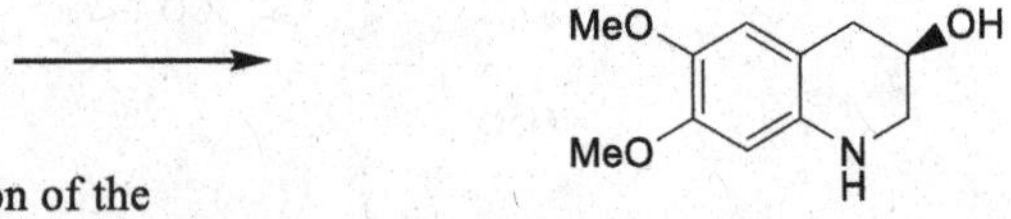

→ Reduction of the imine intermediate F

G   3-(R)

ACYLATION OF THE AMINE, AND INVERSION AMINATION OF THE ALCOHOL

Selective propionylation of the amine in **G** was done with cold propionic anhydride, with retention of chirality of the alcohol at C-3. Inversion of this alcohol in **H**, to the dimethylamine (required in **L**), might have been accomplished with a single Mitsunobu reaction. Instead, a more efficient four-step sequence below was used.

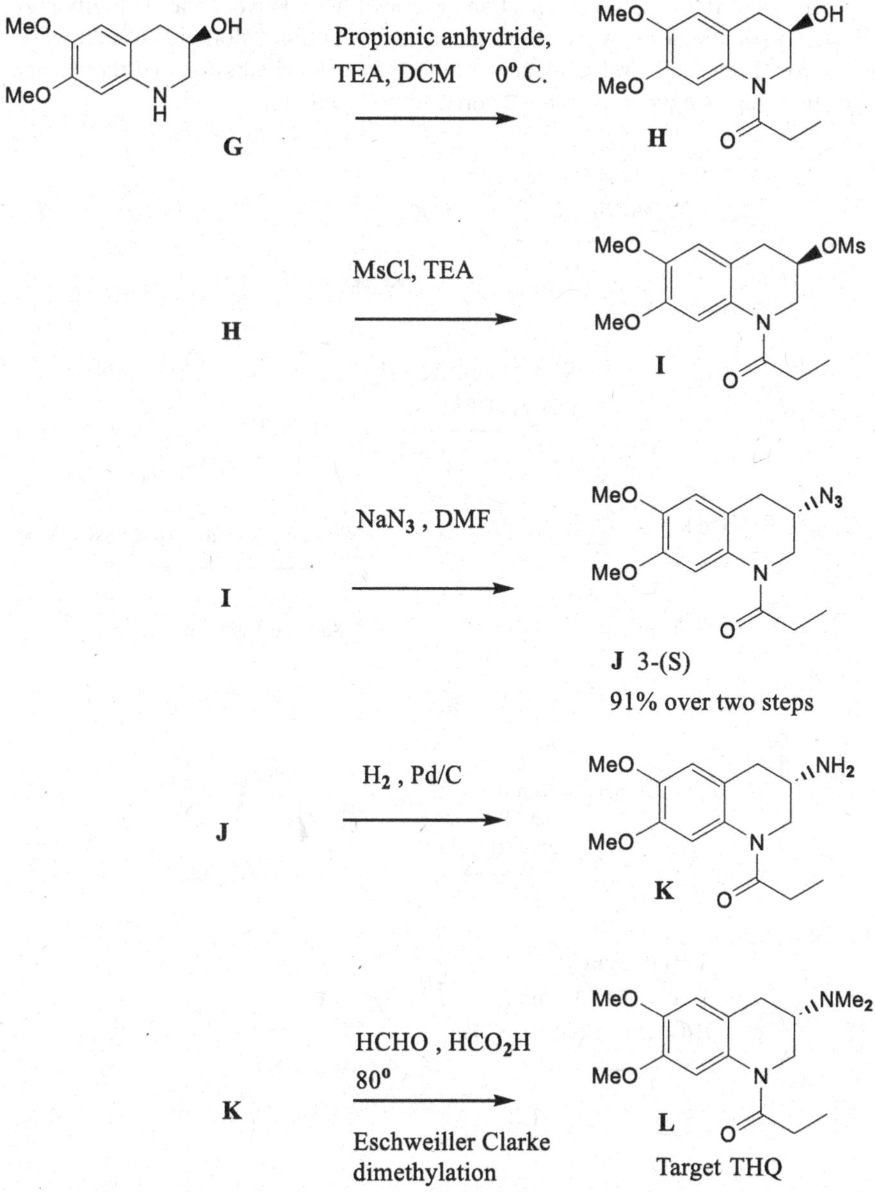

G

Propionic anhydride,
TEA, DCM   0° C.

H

H   MsCl, TEA

I

I   NaN$_3$, DMF

J 3-(S)
91% over two steps

J   H$_2$, Pd/C

K

K   HCHO , HCO$_2$H
80°

Eschweiller Clarke
dimethylation

L
Target THQ

1,3,5-TRIOLMOTIF SYNTHON

Guardiño-Castro, M.H.; Procter, D.J. Diasteroselective hydroxyethylation of β-hydroxyketones: A Reformatsky cyclization–lactone reaction cascade mediated by SmI_2-H_2O. *Helvetica Chim. Acta* **2019**, *102*, el900227

The chiral 1,3,5-trihydroxy grouping occurs in several natural products. The diasteroselective Reformatsky cyclization described here delivers one such grouping of use in their synthesis. A stereospecific synthesis at this Reformatsky reaction (**C** to **D**) would require chiral guidance at both the carbonyl addition and the lactone formation stages in order to obtain a pure (+) triol synthon.

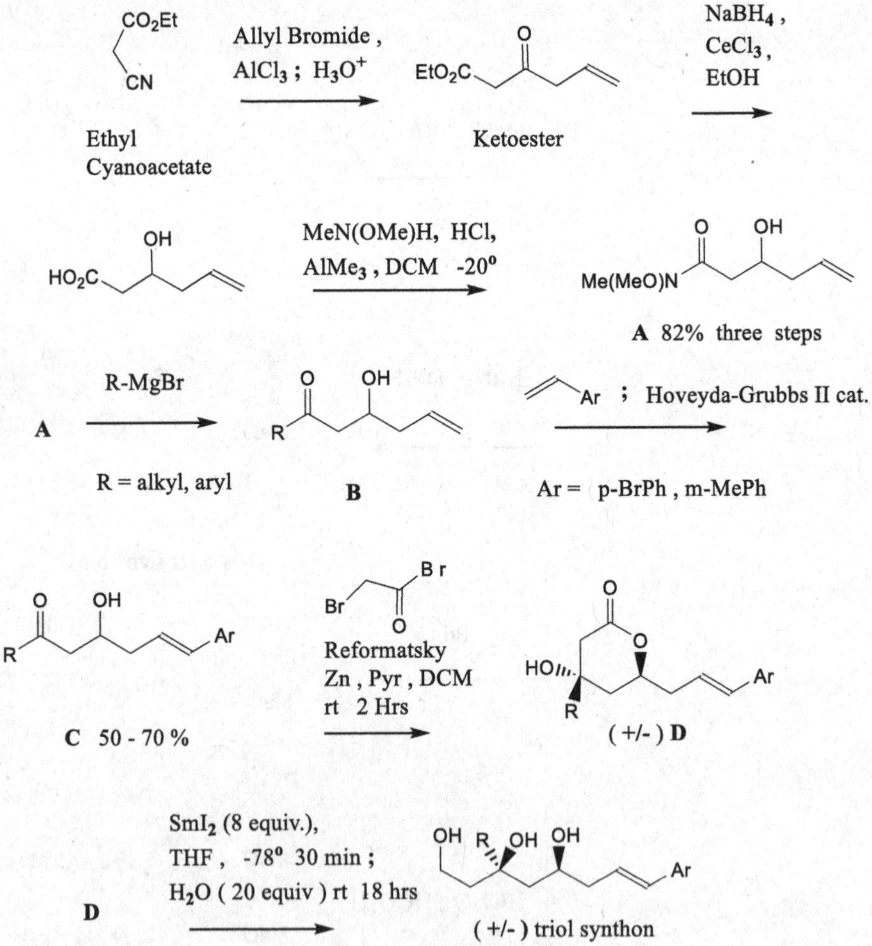

A 82% three steps

B

C 50 - 70 %

(+/-) **D**

(+/-) triol synthon

3 Natural Products

In which more efficient methods are invented for the synthesis of complex compounds. Naturally occurring terpenes, alkaloids, vitamins, and other constituents of herbal medicines have been prepared by synthesis since the early twentieth century, with the objective of producing a drug with therapeutic properties better than the original natural product, using a less expensive and scalable synthesis, and in quantities not available from natural sources. Many natural products can be evaluated for use only if gram quantities of the compounds – not available from natural sources – can be synthesized. Similarly, many natural products can be profitably marketed only if industrial scale processes are possible.

It is in this interest that drug houses, venture capitalists, agricultural corporations, universities, and government agencies continue to sponsor synthesis projects, such as the ones in this and other chapters.

DOI: 10.1201/9781003397816-3

3-EPI JURENOLIDE C

Katsumura, N.; Inagaki, M.; Kiriseko, A.; Saito. Total Synthesis of 3-Epi-Juruenolide C. *Chemical and Pharmaceutical Bulletin* **2019**, *67*, 594–598

Retrosynthetic Plan

The complex furanone ring of Jurenolide C would be prepared from an advanced intermediate such as **I**. For the preparation **of l**, the known compound **G** was to be converted into alkene **H**, then metathesized with sorbate derivative **F**.

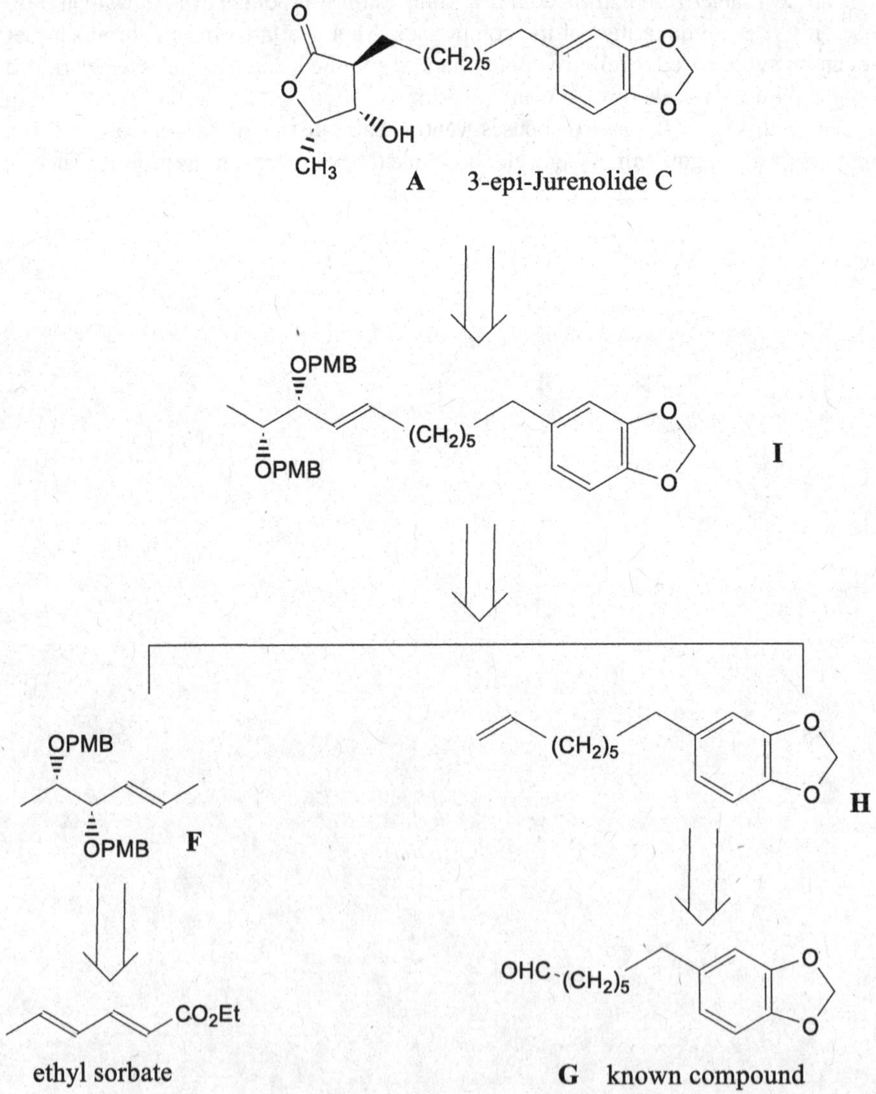

A   3-epi-Jurenolide C

I

F

H

ethyl sorbate

G   known compound

HIGHLIGHTS

- Five step preparation of furanone ring precursor   **Et Sorbate** to **F**
- One pot regioselective bromination   **I** to **J**
- Carbonylation of the bromoalkene with simultaneous
  furanone formation   **L** to **M**
- Remarkable face-selective solvent-dependent
  hydrogenation   **M** to **A**

THE FURANONE RING PRECURSOR

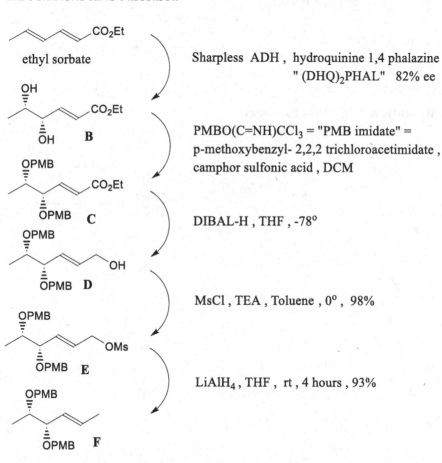

ethyl sorbate

B

C

D

E

F

Sharpless ADH , hydroquinine 1,4 phalazine
" (DHQ)$_2$PHAL"   82% ee

PMBO(C=NH)CCl$_3$ = "PMB imidate" =
p-methoxybenzyl- 2,2,2 trichloroacetimidate ,
camphor sulfonic acid , DCM

DIBAL-H , THF , -78°

MsCl , TEA , Toluene , 0° , 98%

LiAlH$_4$, THF , rt , 4 hours , 93%

LINKER TO BENZODIOXOL

Initial preparation

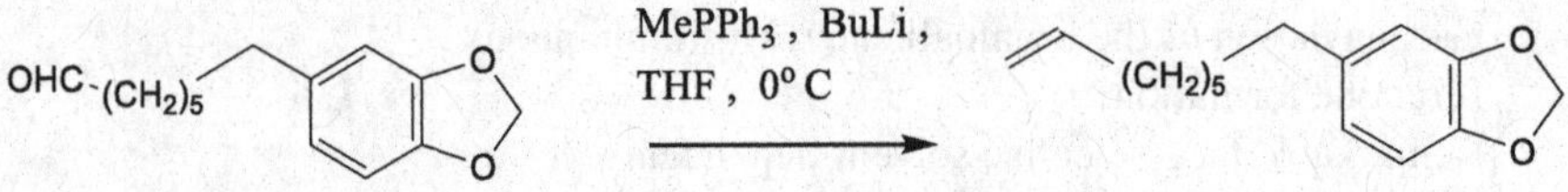

OHC-(CH$_2$)$_5$ [benzodioxole structure]

G   known compound

$\xrightarrow{\text{MePPh}_3 \text{ , BuLi,}}$
THF , 0° C

[alkene-(CH$_2$)$_5$-benzodioxole structure]

H   89%

Alternate Kamuda approach

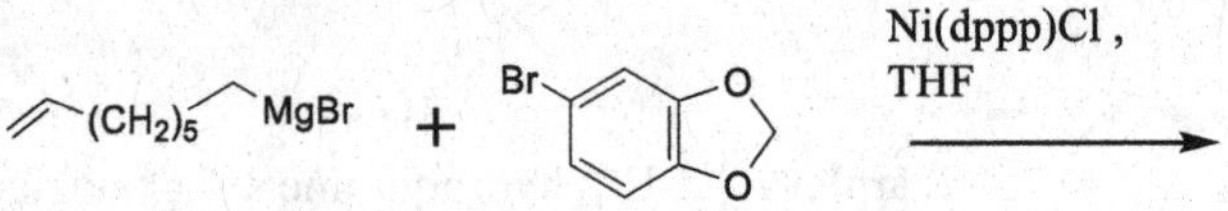

[alkene-(CH$_2$)$_5$-MgBr structure]   **+**   Br-[benzodioxole structure]

$\xrightarrow{\text{Ni(dppp)Cl ,}}$
THF

H 29%

METATHESIS WITH LACTONE PRECURSOR

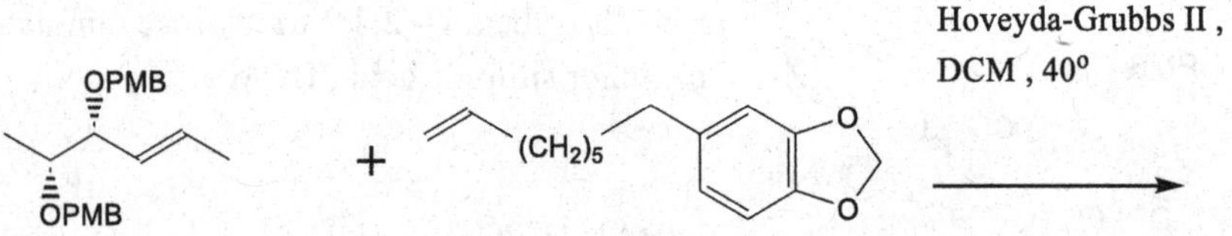

[structure with OPMB, OPMB groups]

F

+

[alkene-(CH$_2$)$_5$-benzodioxole structure]

H

$\xrightarrow{\text{Hoveyda-Grubbs II ,}}$
DCM , 40°

[product structure with OPMB, OPMB, (CH$_2$)$_5$ and benzodioxole]

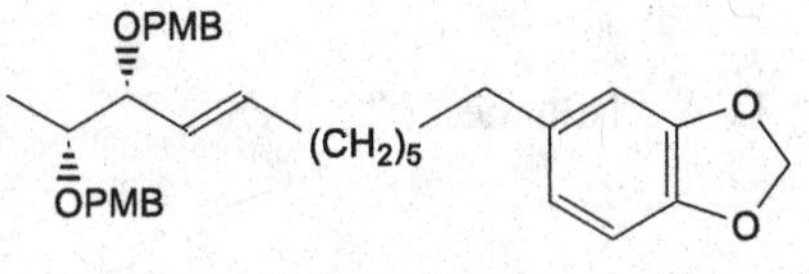

I   65%

dppp = 1,3 bis(diphenylphonphino) propane = Ph$_3$P(CH$_2$)$_3$PPh$_3$

BROMINATION AND CARBONYLATION, WITH SPONTANEOUS LACTONE FORMATION

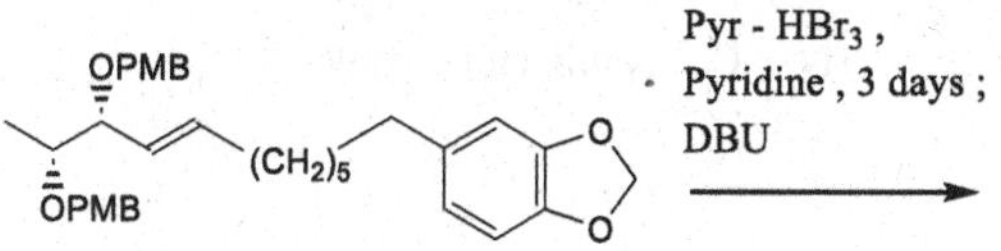

OPMB

ŌPMB

(CH₂)₅

I

Pyr - HBr₃ ,
Pyridine , 3 days ;
DBU

Reaction product is a multicomponent
mixture of monobromides, including
the two shown below as **J** and **K**

PMBO   Br

PMBŌ   (CH₂)₅

H

J

+

PMBO   H

PMBŌ   (CH₂)₅

Br

K

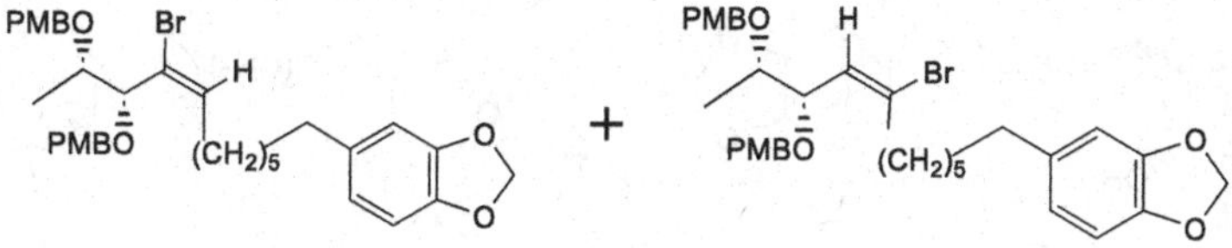

J + K
Mixture

6M HCl ,
reflux

Deprotection

OH   Br

ŌH   (CH₂)₅

H

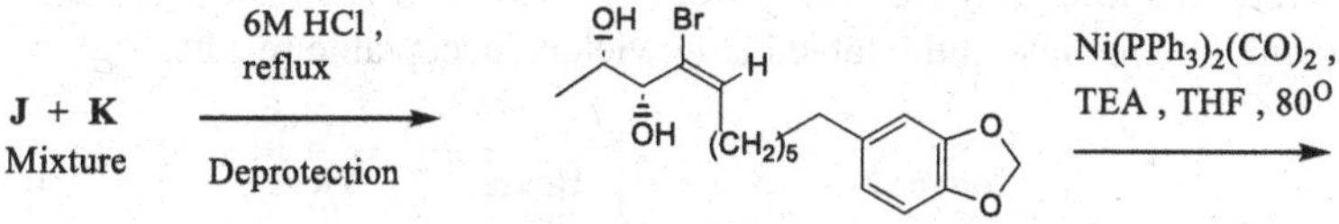

L

67% of this single
deprotection product.

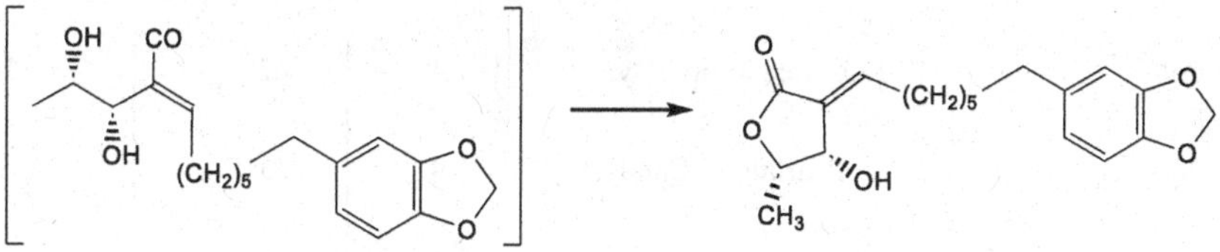

Ni(PPh₃)₂(CO)₂ ,
TEA , THF , 80°

OH   CO

ŌH   (CH₂)₅

(CH₂)₅

CH₃

⁰OH

M

ENANTIOSELECTIVE HYDROGENATION

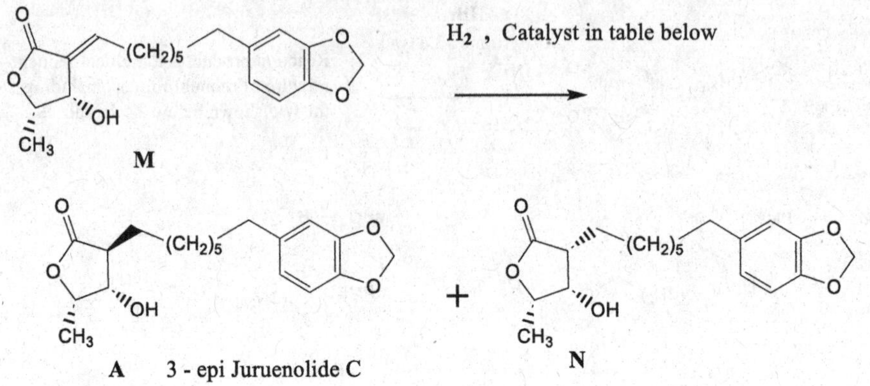

H_2 , Catalyst in table below

M

A   3 - epi Juruenolide C

N

Products were quantitated by reverse-phase HPLC. Of all the conditions screened for hydrogenation, the last in the table below yielded acceptable results.

Catalyst	Comments	Hours	Yield	Selectivity
			$(A + N)\%$	$(A : N)$
Pd/C	Expected to give Syn N	4	62	1 : 20
$Pd(OH)_2$	Pearlman's Catalyst	4	76	1 : 19
$Ir(cod)(pyr)(PPh_3)^+PF_6^-$	Crabtree's Catalyst	19	70	3 : 2
$Rh(PPh_3)_3^+Cl^-$	Wilkenson's Catalyst Solvent dependance :			
	toluene	20	70	1 : 1
	ethyl acetate	19	70	2 : 3
optimal conditions	dichloromethane	24	72	5 : 1

MORPHINE

Brousseau, J.; Xolin, A.; Barriault, L.A. A Nine-Step
Formal Synthesis of (+/–)-Morphine. *Org. Lett.* **2019**, *21*,
1347–1349

Highlights

- Diels–Alder reaction with neat Danashefski's Diene (**A** to **B**)
- Vinyl ether preparation using ethyl vinyl ketone, with prompt Claisen cascade that alkylates the atomatic ring (**B** to **C**)
- Oxygenation at an allylic carbon followed immediately with stereoselective reduction of that ketone and reduction of an ester to the aldehyde (**C** to **D**)
- An uncommon alkene hydroamination (**F** to morphine)

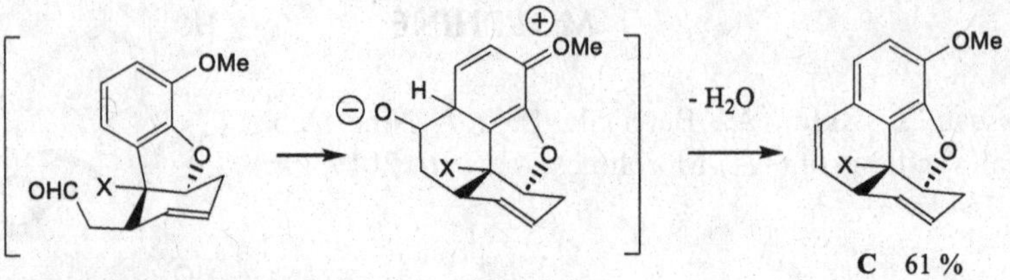

(The Claisen rearrangement is followed by arylation
of the aldehyde by the electron-rich benzene ring,
then elimination of water)

C   61 %

Phenanthrofuran
Core

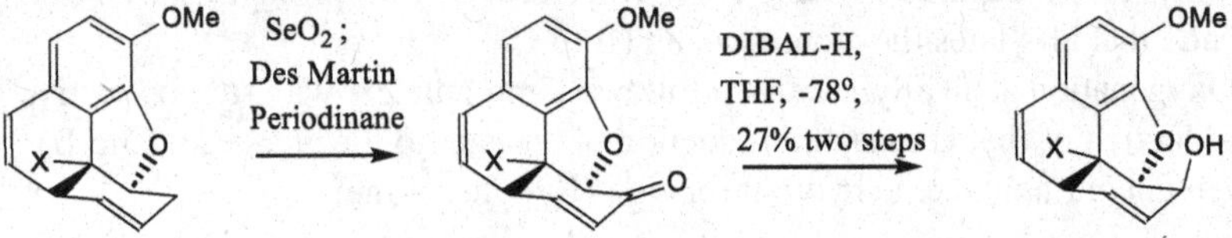

C

X = CO$_2$Et
(unstable with
with base, or on
silica gel)

D   X = -CHO

The crude ketoester is promptly reduced
under selective conditions that converts
the ester to the aldehyde (without furthur
reduction) and the ketone to the alcohol.

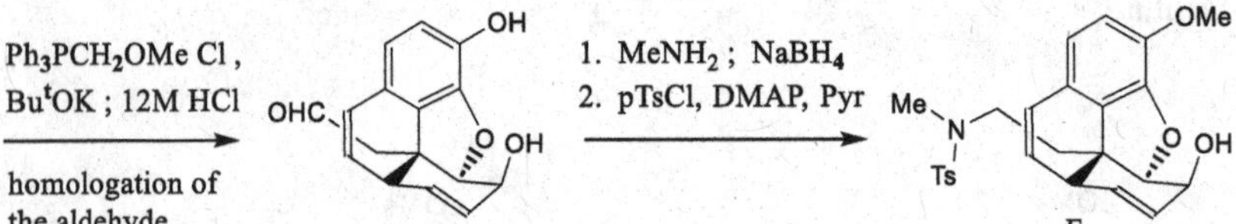

Ph$_3$PCH$_2$OMe Cl ,
Bu^{t}OK ; 12M HCl

homologation of
the aldehyde

E

1. MeNH$_2$; NaBH$_4$
2. pTsCl, DMAP, Pyr

F

Li , NH$_3$, THF ,
Bu^{t}OH , -78^0

60%
Guilliou's
hydroamination
procedure

Morphine

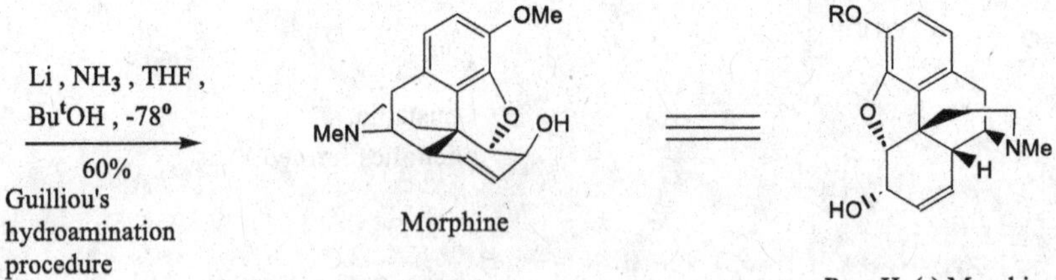

R = -H (-) Morphine
R = -CH$_3$ (-) Codeine

TOCOPHEROL

Tsubogo, T.; Aoyama, S.; Takeda, R.; Uchiro, H. Synthesis of 2,2-dialkyl chromans by intramolecular Ullmann C-O coupling reactions toward the total synthesis of D-α-tocopherol. *Chemical and Pharmaceutical Bulletin* **2018**, *66*, 843–846

This synthesis of tocopherol was guided by the retrosynthetic plan below. Early in the synthesis the phytyl chain in **A** is linked with chiral epoxide **B**, establishing the chiral center in aldehyde **F**. Linkage with the totally substituted aromatic Wittig reagent **G′** produces iodoalcohol **I**, lying an Ullmann coupling away from the target **K**.

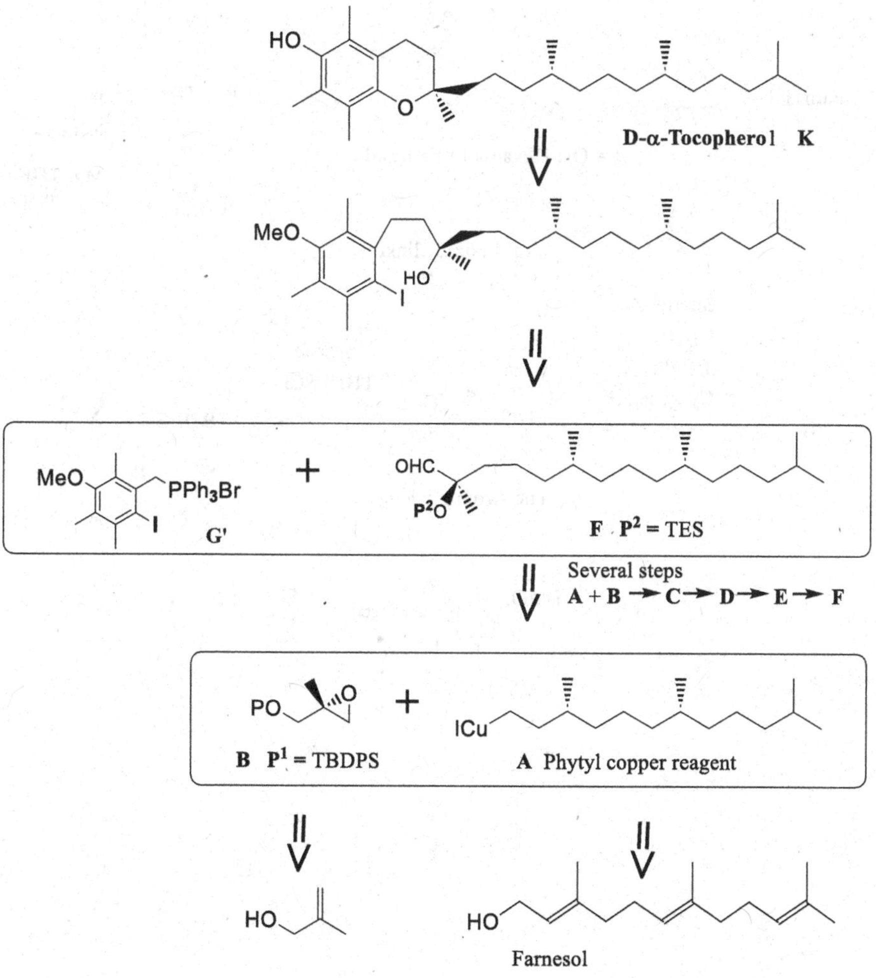
HO
O
D-α-Tocopherol K
MeO
HO
I
MeO
PPh3Br
I
G'
+
OHC
P²O
F P² = TES
Several steps
A + B → C → D → E → F
PO
O
+
ICu
B P¹ = TBDPS
A Phytyl copper reagent
HO
HO
Farnesol

D-A-TocopheroI K

STARTING MATERIALS

Phytyl Chain

Farnesol $\xrightarrow{\text{H}_2 \text{ / Cat.}}$ R

R = OH Hexahydro Farnesol

$R = OH$ ⎫ Ph₃P
⎭ NBS

$R = Br$

A $R = CuI$ Mg, THF ;
CuI, THF

Chiral Epoxide linker

Sharpless AE
(-)-DIPT
Ti(OPr)₄
Cymene HP

HO $\xrightarrow{\hspace{2cm}}$ HO $\xrightarrow{\text{TBDPSCl}}$ TBDPSO **B**

The Aromatic Ring

MeO PPh₃Br
I **G'**

Attachment of the Phytyl Chain to the Chiral Linker, and Then to the Aromatic Ring

Unconventional Reduction of the Double Bond in G, Then the Title UllmanIntramolecularnand equally Unconventional Demethylation to Tocopherol

Demethylation of the phenolic methyl ether is realized with an alternative to the boron trihalides.

ACREMINES F, A, AND B

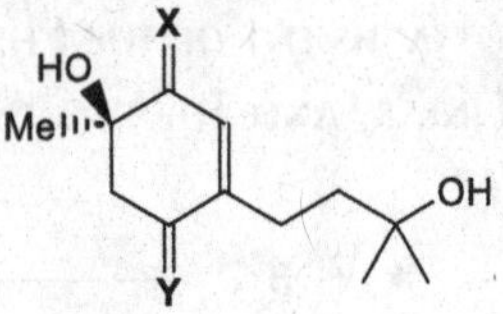

Winter, N.; Trauner, D. Synthesis of acremines
A, B, and F and studies on the bisacremines.
Beilstein J Org. Chem. **2019**, *15*, 2271–2276.

The Acremenes

$$X \text{ or } Y = \begin{smallmatrix} OH \\ H \end{smallmatrix} \text{ or } =O$$

The Acremines are meroterpenoids from a fungus of the genus *Acremonium*. A modest amount of these compounds are needed to study the bioactivity of trace amounts isolated from the discovery natural product. The first synthesis of Acremine F begins with the preparation of key intermediate I, in low yield but high purity, as shown below.

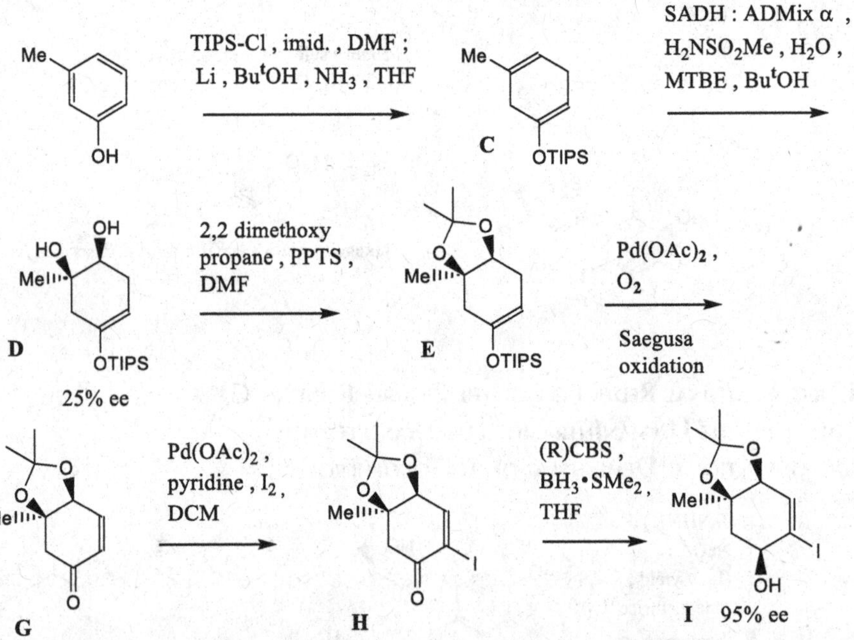

TIPS-Cl , imid. , DMF ;
Li , Bu^{t}OH , NH$_3$, THF

SADH : ADMix α ,
H$_2$NSO$_2$Me , H$_2$O ,
MTBE , Bu^{t}OH

C

D   25% ee

2,2 dimethoxy
propane , PPTS ,
DMF

E

Pd(OAc)$_2$,
O$_2$

Saegusa
oxidation

G

Pd(OAc)$_2$,
pyridine , I$_2$,
DCM

H

(R)CBS ,
BH$_3$·SMe$_2$,
THF

I   95% ee

The enantioselective Sharpless assymmetric dihydroxylation (**SADH**) of enolate **C** gave the product **D** with the desired regiospecificity, but in poor yield and enantioselectivity. As this crude product was carried through acetonide **E** formation, oxidation to enone **G**, then to α-iodoenone **H**, this racemic mix was retained. In the course of the Corey–Itsuno reduction of **H**, however, the desired intermediate I was obtained in low yield, but with surprisingly good enantiomeric excess.

(R)CBS =
Corey Bakshi Shibata
reagent.

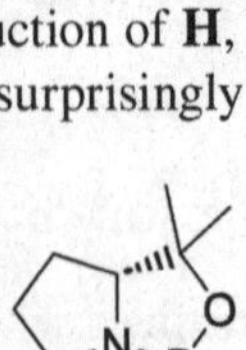

TIPS-Cl = triisopropylsilyl chloride
SADH = Sharpless assymmetric dihydroxylation
PPTS = pyridinium p-toluene sulfonate
MTBE = methyl *tert*-butyl ether

Deprotection of iodo-acetonide **I** and Stille coupling with **K** delivered Acremine F.

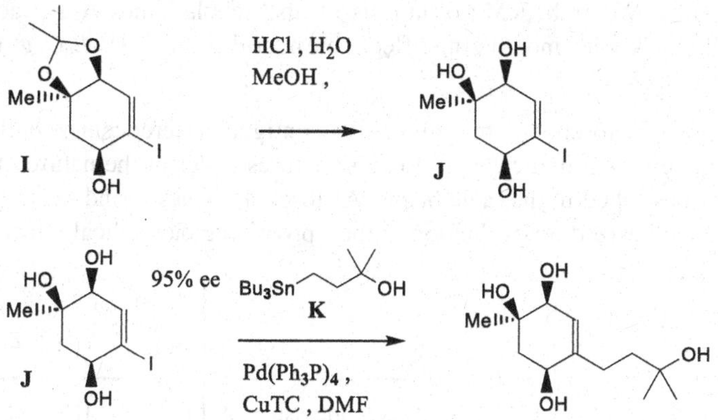

Acremine F

The allylic hydroxyl groups of Acremine F itself were oxidized selectively to Acremines A and B by using different stoichiometric amounts of iodoxybenzoic acid "IBX." Oxidizing agents that exhibit such selectivity, as shown below for these two similar allylic alcohols, are useful in the course of many syntheses.

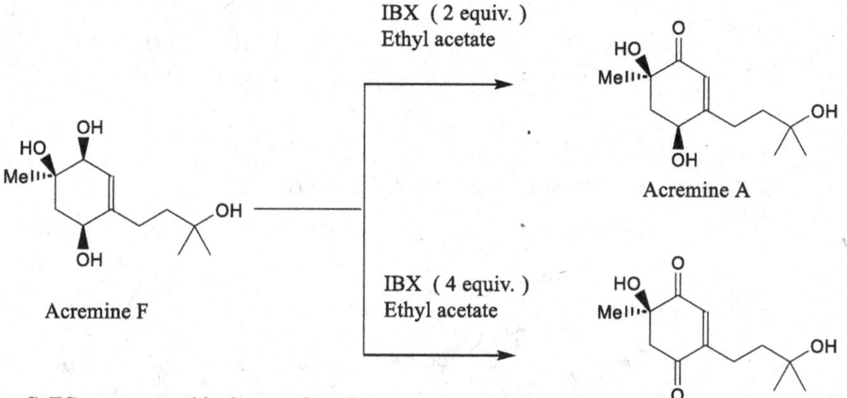

IBX (2 equiv.)
Ethyl acetate

Acremine A

IBX (4 equiv.)
Ethyl acetate

Acremene B

CuTC = cuprous thiophene carboxylate

ALTOSCHOLARSINE C AND D

Mason, J.D.; Weinreb, S.M.; Synthesis of alstoscholarisines A–E, monoterpene alkaloids with modulating effects on neural stem cells. *J. Org. Chem.* **2018**, *83*, 5877

The five known alstoscholarisines all exhibit contiguous stereocenters embedded in a pentacyclic ether-aminal cage. The five structures differ in the nature of the substituents, as described in the table below. Alstoscholarisines C and D are the last of these to be synthesized for evaluation of their promising biomedical effects.

	X	Y	Z
Alstoscholarisine A	CH$_3$	H	H
B	CH$_3$	H	CO$_2$Me
C	H	CH$_3$	CO$_2$Me
D	H	CH$_3$	CO$_2$H
E	H	CH$_3$	H

SYNTHETIC OUTLINE OF THE SUCCESSIVE RING CLOSINGS

Alstoscholarsine C

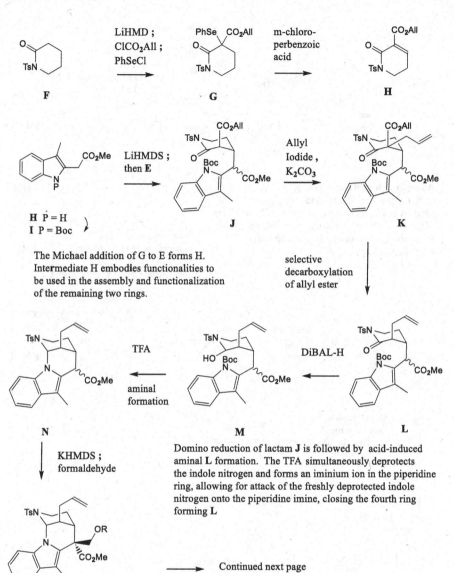

F

G

H

H P = H
I P = Boc

J

K

The Michael addition of G to E forms H. Intermediate H embodies functionalities to be used in the assembly and functionalization of the remaining two rings.

selective decarboxylation of allyl ester

N

M

L

Domino reduction of lactam **J** is followed by acid-induced aminal **L** formation. The TFA simultaneously deprotects the indole nitrogen and forms an iminium ion in the piperidine ring, allowing for attack of the freshly deprotected indole nitrogen onto the piperidine imine, closing the fourth ring forming **L**

Continued next page

O R = H
P R = TBS

TBSOTf, 2.6-lutidine

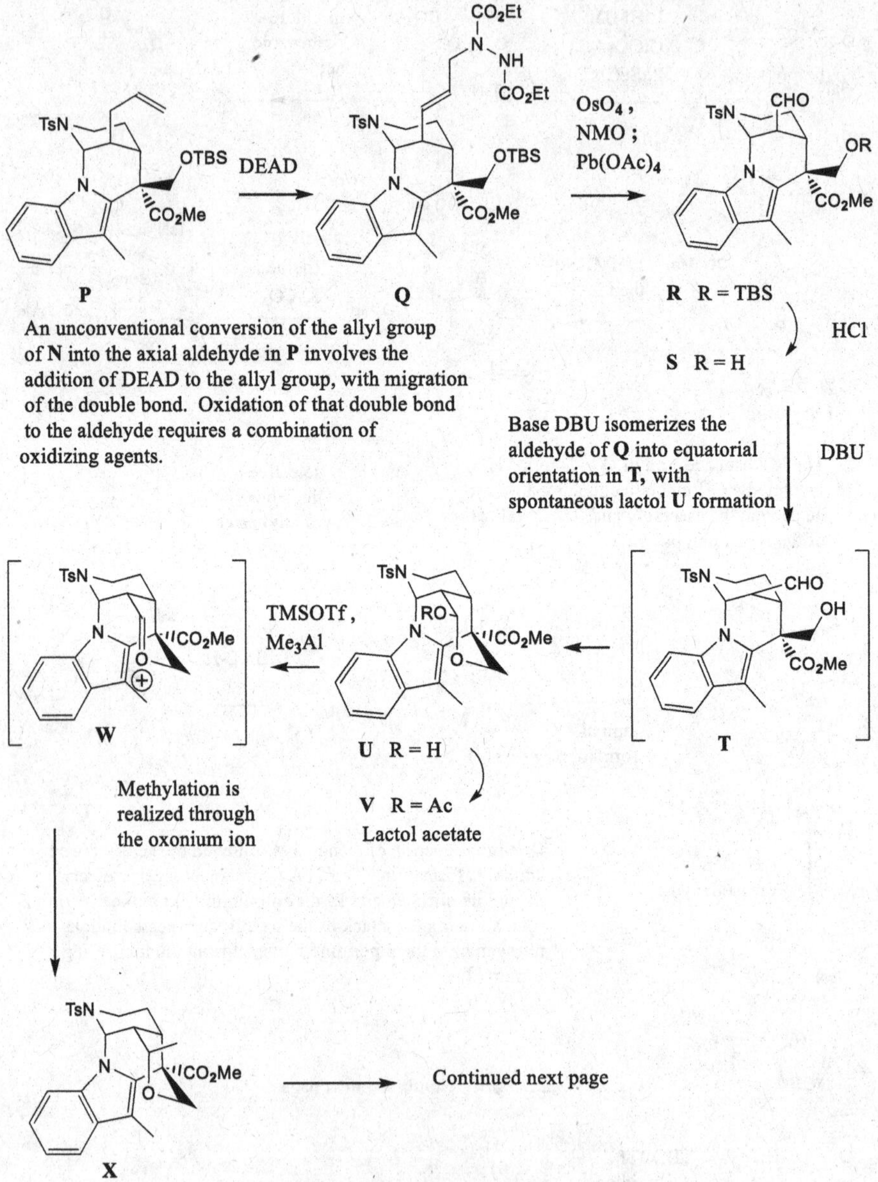
CO2Et
N
NH
CO2Et
TsN
OTBS
N
CO2Me
DEAD
OsO4,
NMO;
Pb(OAc)4
CHO
OR
P
Q
R R = TBS
S R = H
HCl
An unconventional conversion of the allyl group of N into the axial aldehyde in P involves the addition of DEAD to the allyl group, with migration of the double bond. Oxidation of that double bond to the aldehyde requires a combination of oxidizing agents.
Base DBU isomerizes the aldehyde of Q into equatorial orientation in T, with spontaneous lactol U formation
DBU
OH
T
RO
O
TMSOTf,
Me3Al
W
U R = H
V R = Ac
Lactol acetate
Methylation is realized through the oxonium ion
X
Continued next page

DEAD = diethyl azodicarboxylate

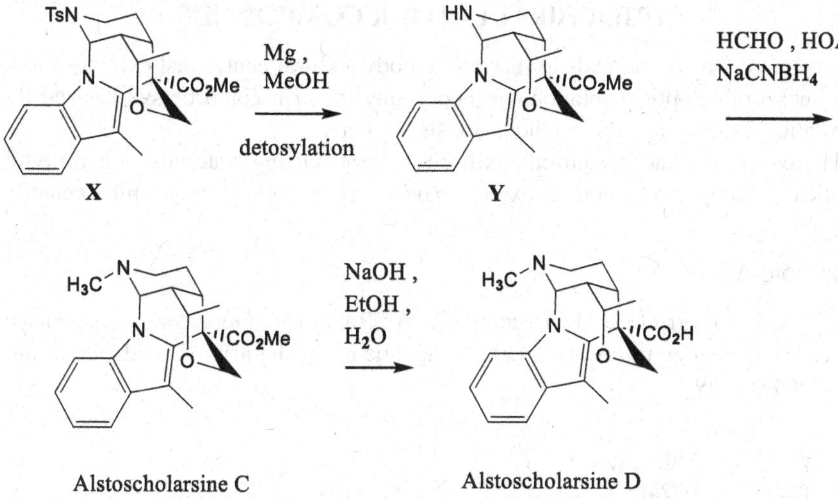

X

Y

Mg ,
MeOH

detosylation

HCHO , HOAc ,
NaCNBH$_4$

Alstoscholarsine C

NaOH ,
EtOH ,
H$_2$O

Alstoscholarsine D

CRYPTOCHIRAL I – FOUR COMPOUNDS

Some compounds of biomedical interest embody a chiral center that exhibits a low or unmeasurable optical rotation. In theory, any of them could be synthesized by conventional methods. Four methods are shown here.

These reports show how chiral auxiliaries, chiral starting materials, or hydrolytic kinetic resolution of a racemic epoxide can be used to generate remote chiral centers.

I Arunoic Acid

Garcia, J.M.; Odriozola, J.M.; Lecumberri, A.; Razkin, J.; Gonzalez, A. A concise and efficient route to the Alzheimer's therapeutic agent (R)-arunoic acid. *Tet.* **2008**, *64*, 10664–10669

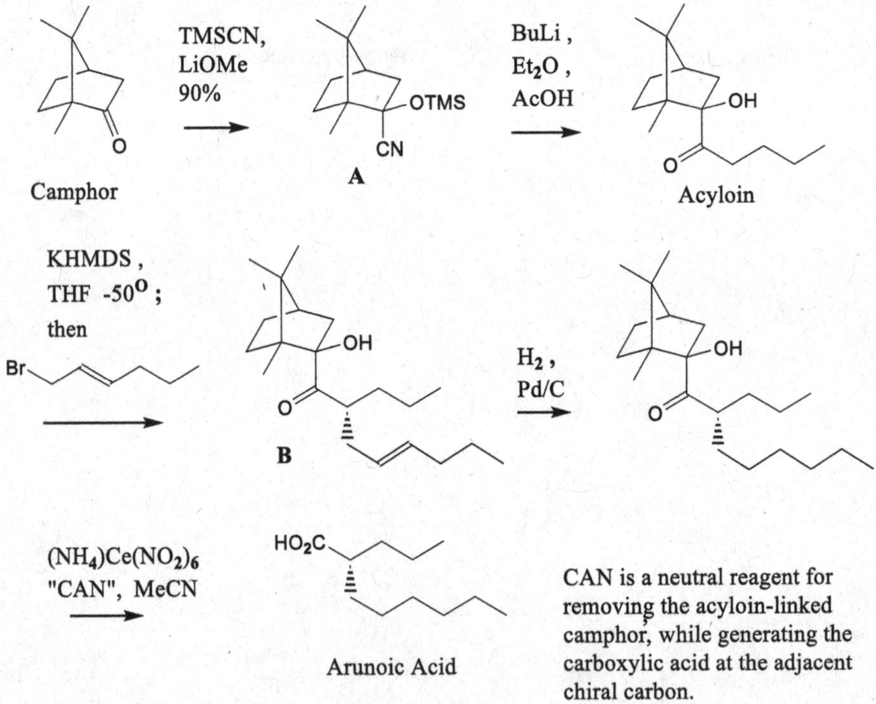

O
Camphor
TMSCN,
LiOMe
90%
OTMS
CN
A
BuLi ,
Et₂O ,
AcOH
OH
O
Acyloin
KHMDS ,
THF -50° ;
then
Br
OH
O
B
H₂ ,
Pd/C
OH
O
(NH₄)Ce(NO₂)₆
"CAN", MeCN
HO₂C
Arunoic Acid
CAN is a neutral reagent for
removing the acyloin-linked
camphor, while generating the
carboxylic acid at the adjacent
chiral carbon.

II Hachijodine B 1: 1-(3-Pyridino)- (S)10-Methyl-12-Amino Dodecane

Romeril, S.O.; Lee, V.; Baldwin, J.E. The synthesis of hachijodine B 1 and a comment on the structure of ikimine B and on the absolute configuration of niphatesine D. *Tet. Lett.* **2004**, *45*, 3273–3277

Readily available chiral terpenes like citronellol are favored starting materials for the preparation of building blocks used in pharmaceutical manufacture.

G Lipophilic Cryptochiral Diamine Target : Hachijodine B 1

TBDMSCl = *tert*-butyl dimethylsilyl chloride
DMPU   = N, N'-dimethylpropylene urea
imid   = imidazole

III HYDROLYTIC KINETIC RESOLUTION (HKR): (R) PENTADEC-1-ENE-4-OL

Harbindu, A.; Sharma, B.M.; Kumar, P. Asymmetric routes to pentadec-l-ene-4-ol: application to the synthesis of aculeatins F and epi-F, (R)- and (S)-5-hexadecanolide and a formal synthesis of solenopsin. *Tet. Asym.* **2013**, *24*, 305–314

Catalytic hydrolysis of racemic epoxide **J** produces an equal amount of resolved epoxide (R) **K** and a chiral diol **L**. This inefficiency is balanced by the overall practicality of this synthesis: and the unneeded diol **L** could be recycled to **K**.

M Pentadec-1-ene-4-ol

This compound has been used to make a variety of cryptochiral compounds by metathesis with other alkenes, followed by hydrogenation of the double bond.

IV 4(R)-METHYLOCTANOL

Muñoz, L.; Pilar Bosch, M.; Rosell, G.; Guerrero, A. Asymmetric synthesis of (R)- and (S)-4-methyloctanoic acids. A new route to chiral fatty acids with remote stereocenters. *Tet. Asymm.* **2009**, *20*, 420–424

(S,S) N

Pseudoephedrine amide
of propionic acid

The (R,R) and (S,S) diastereomers of pseudo-ephedrine propionamide (**N**) were exploited to prepare (R) 4-methyloctanol (**U**).

Removal of the chiral auxilliary (**0** to **P**) was realized with an unconventional reducing agent after previously favored reagents – such as LiAlH(OEt)3 or TFA – failed.

CRYPTOCHIRAL II: LEAFMINER PHERMONE

Taguri, T.; Yamakawa, R.; Fujii, T.; Muraki, Y.; Ando, T. Stereospecific inversion of secondary tosylates to yield chiral methyl-branched building blocks, applied to the asymmetric synthesis of leafminer sex pheromones. *Tet. Asymm.* **2012**, *23*, 852–858

RETROSYNTHETIC GUIDELINE

The near symmetry of the target **K** dictates a strategic disconnect (◌◌◌) at the C7–C8 bond that led to symmetric synthetic sequences for the Western (**F**) and Eastern fragments (**L**) in the guideline shown below.

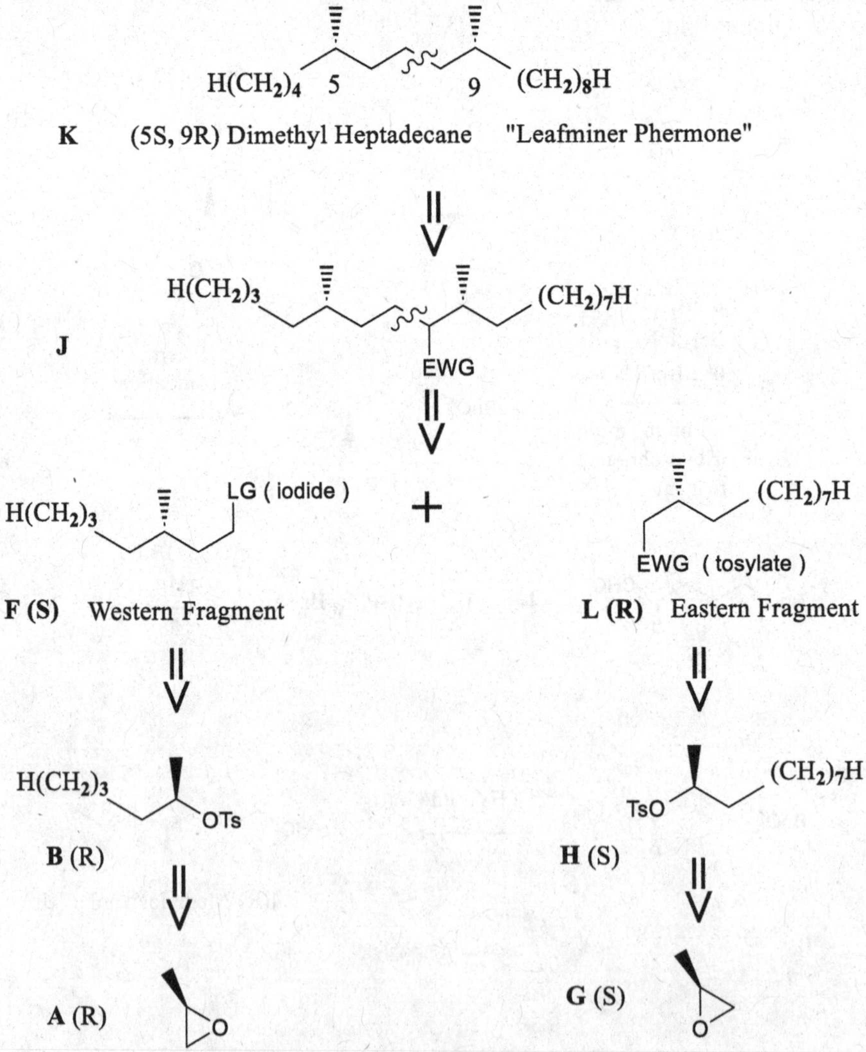

H(CH$_2$)$_4$ 5   9   (CH$_2$)$_8$H

K   (5S, 9R) Dimethyl Heptadecane   "Leafminer Phermone"

H(CH$_2$)$_3$ (CH$_2$)$_7$H

J

EWG

H(CH$_2$)$_3$ LG (iodide)   +   (CH$_2$)$_7$H

EWG (tosylate)

F (S)   Western Fragment   **L (R)**   Eastern Fragment

H(CH$_2$)$_3$ (CH$_2$)$_7$H

OTs   TsO

B (R)   **H (S)**

A (R)   O   **G (S)**   O

Western C_7 Fragment (F)

A (R) → [H(CH$_2$)$_3$MgCl, Li$_2$CuCl$_5$; TsCl] → B (R) [H(CH$_2$)$_3$ ... OTs] → [diethyl malonate, DME (dimethoxyethane)]

C (S) [H(CH$_2$)$_4$ ... CO$_2$Et, CO$_2$Et] → [LiCl, DMSO 180°] → D (S) [H(CH$_2$)$_4$ ... CO$_2$Et]

D (S) → [LAH] → [H(CH$_2$)$_4$ ... X]

E (S) X = OH
F (S) X = Iodine
[I$_2$, Ph$_3$P, Imidazole]

This example has three alkyl substituents on the chiral carbon. However, this common construction of a chiral center has been successful where a wider variety of functional substituents on the chiral carbons have been involved.

Eastern C10 Fragment (L)

G (S) → [H(CH$_2$)$_7$MgBr; TsCl] → H (S) [TsO ... (CH$_2$)$_7$H]

H (S) + [tosyl methyl sulfone] → [BuLi, THF 55%] → L (R) [... (CH$_2$)$_7$H, SO$_2$]

The tosyl methyl sulfone serves in **L** to activate the adjacent methyl group for coupling with the Western C_7 fragment **F** in the next sequence and is also conveniently removable.

WESTERN–EASTERN COUPLING (F + L)

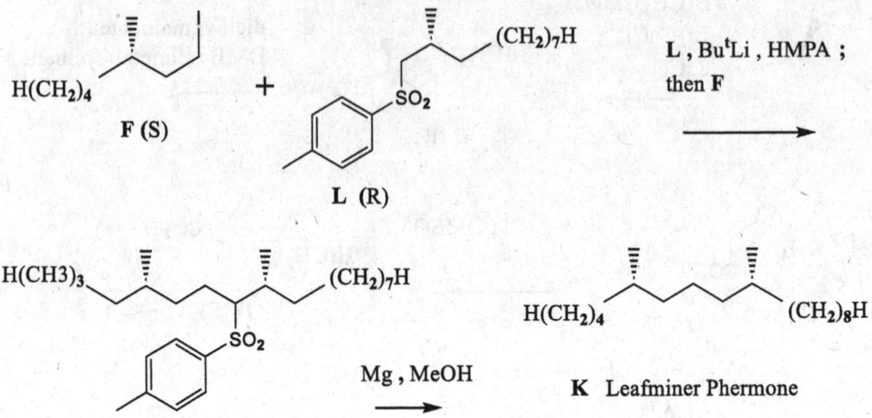

F (S)

+

L (R)

L, Bu^{t}Li, HMPA ;
then F

J (5S , 9R)

Mg , MeOH

K   Leafminer Phermone

The magnesium/methanol reagent used for detosylation (**J** to **K** above) has been found to reduce other functionalities

OXYPALMATINE

Gadhiya,S.; Ponnala, S.; Harding,W.W. A divergent route to 9.10-oxygenated tetrahydroprotoberberine and 8-oxoprotoberberine alkaloids: synthesis of (+/-)-isocorypalmine and oxypalmatine. *Tetrahedron* **2015**, *71*, 1227–1231

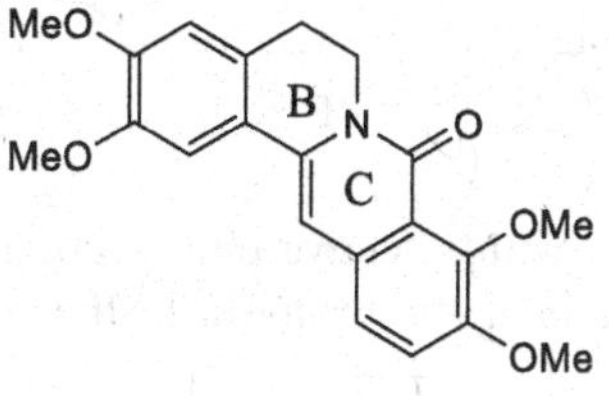
MeO
MeO
B
C
N
O
OMe
OMe

Oxypalmatine

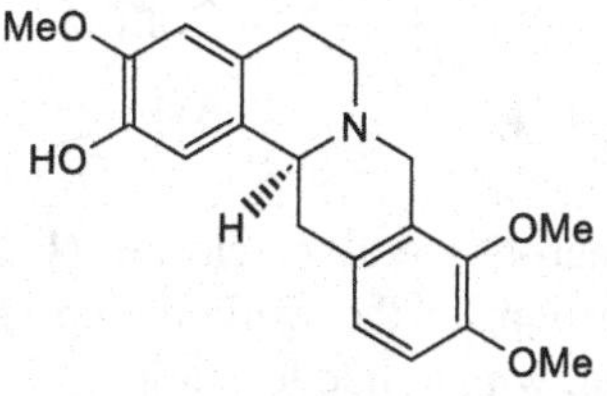
MeO
HO
N
H
OMe
OMe

(-) Isocorypalmine

An uncommon malonate alkylation of an intermediate benzyne, shown below with rearrangement, provided the scaffold **C** for the closure of both Band C rings. A two-stage one-pot reaction (**E** to **G**) delivers the tetracyclic core of oxypalmatine.

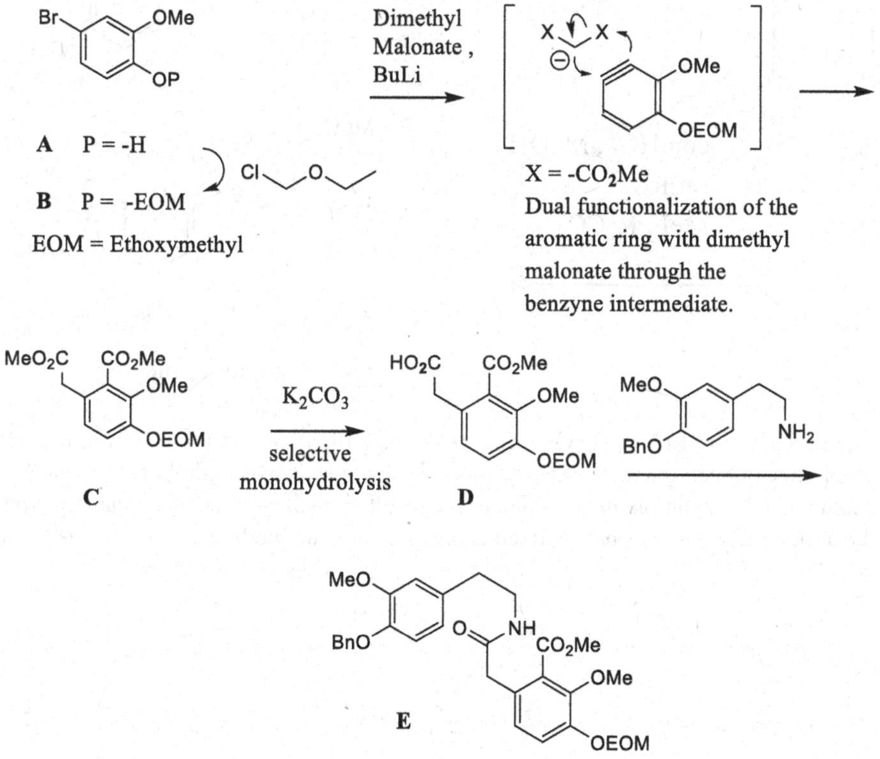
Br
OMe
OP
Dimethyl Malonate , BuLi
X
X
OMe
OEOM
A P = -H
Cl O
B P = -EOM
EOM = Ethoxymethyl
X = -CO2Me
Dual functionalization of the aromatic ring with dimethyl malonate through the benzyne intermediate.
MeO2C
CO2Me
OMe
OEOM
C
K2CO3
selective monohydrolysis
HO2C
CO2Me
OMe
OEOM
D
MeO
BnO
NH2
MeO
BnO
O
NH
CO2Me
OMe
OEOM
E

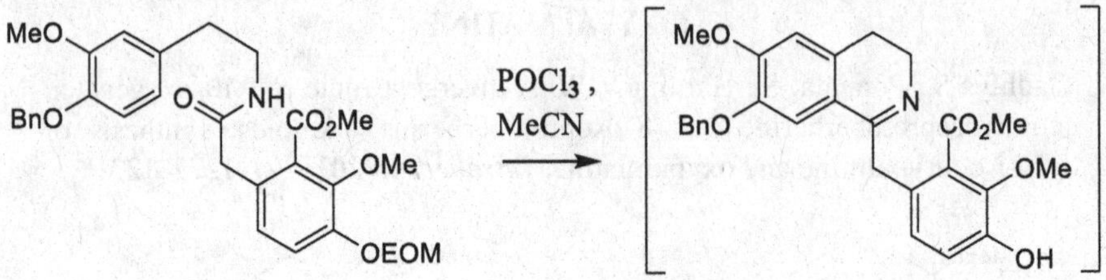

E   F

Bishler–Napieralsky closure (**E** to **F**) occurs with selective deprotection of the EOM group. In the workup step (**F** to **G**), the benzylic proton a to the imine is removed, with amide formation.

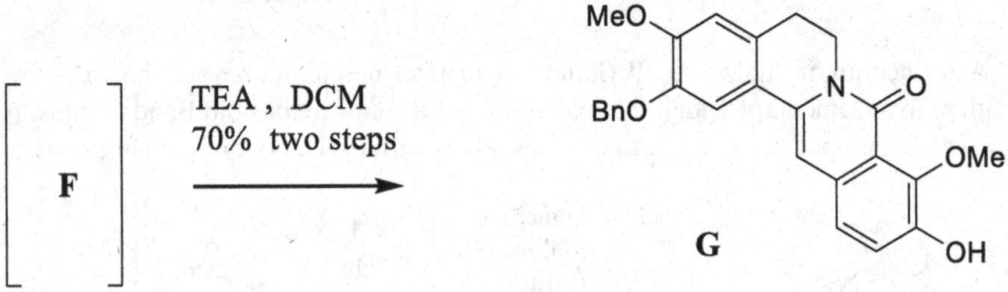

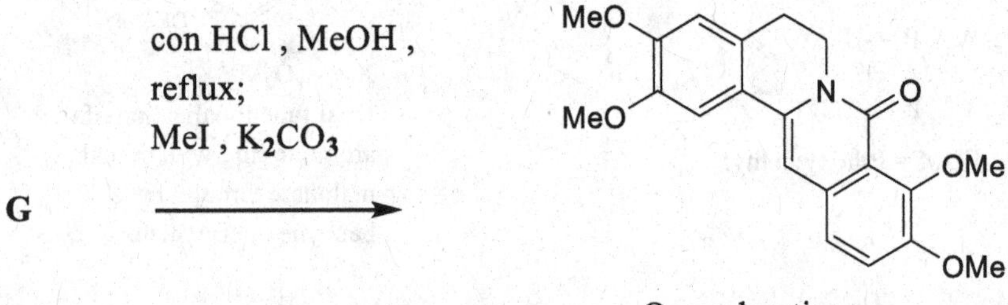

TEA , DCM
70% two steps

F   **G**

con HCl , MeOH ,
reflux;
MeI , K$_2$CO$_3$

G

Oxypalmatine

The preparation of (-) isocorypalmine from oxypalmatine had not been optimized as of this publication date. Such a required face-selective hydrogenation of the enamide, debenzylation, methylation of the resultant hydroxyl, and deoxygenation of the amide carbonyl, was not realized using conventional methods.

GALEILLALACTONE

(-) Galiellalactone

Kim, T.; Han, Y.T.; Kim, K.; Lee, J.; Suh, Y.-G. Diastereoselective Total Synthesis of(-) Galiellalactone. *J Org. Chem.* **2015**, *80*, 12193–12200

Kim, H.S.; Kim, T.; Ko, H.; Lee, J.; Suh, Y.G. Identification of galiellalactone-based novel STAT3-selective inhibitors with cytotoxic activity against triple-negative breast cancer cell lines. *Bioorg. Med. Chem.* **2017**, 25, 5032–5040

HIGHLIGHTS

- An unsymmetrically substituted silyl protecting group (**B** to **C**)
- An intramolecular Tsuji-Trost cyclization, facilitated using a sacrificial sulfinate group (**G** to **H**)
- Three successive and selective oxidations of a triol and an allylic alcohol (**K** to **N**)
- Lewis acid-selective enantiodiscriminant Hosomi–Sakurai reaction (**N** to **O**)
- A modified Barton-McCombie deoxygenation of a pivotal hydroxyl group (**R** to galiellalactone)

PMB = *p*-methoxybenzyl
TBAF = tetrabutyl ammonium chloride
TBSCl = *tert*-butylsilyl chloride
DCM = dichloromethane

complexation and cyclization
to **H** and **I** on next page

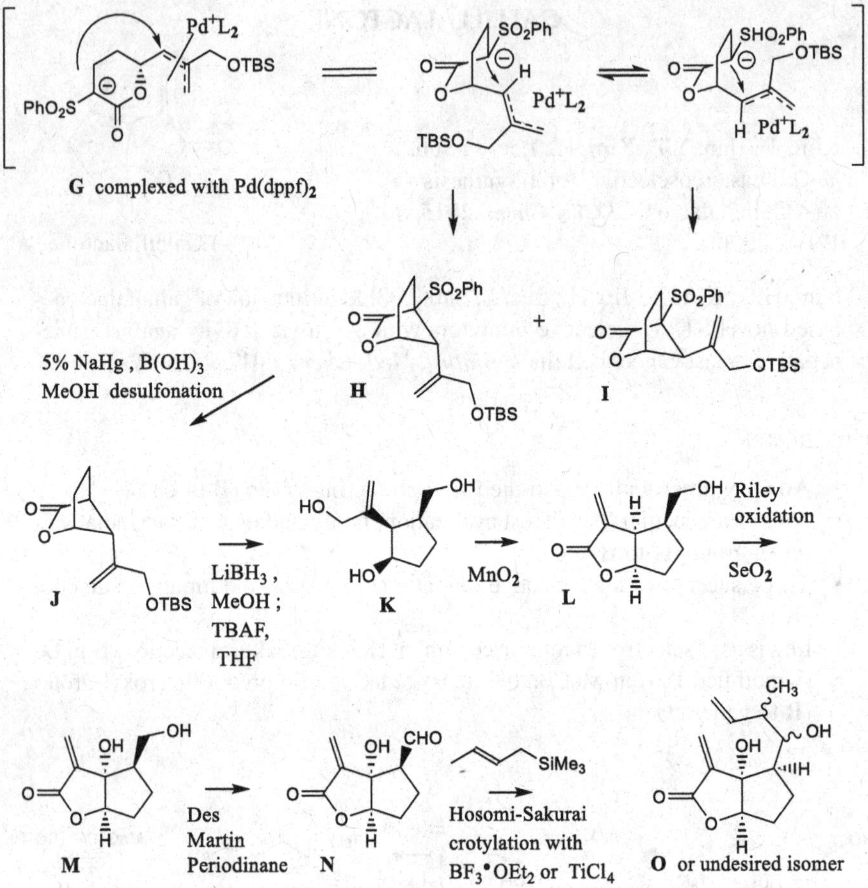

G complexed with Pd(dppf)$_2$

5% NaHg , B(OH)$_3$
MeOH desulfonation

H   +   I

J   LiBH$_3$,
OTBS   MeOH ;
          TBAF,
          THF

K   MnO$_2$ L   Riley
                                        oxidation

                                        SeO$_2$

M   Des   N   Hosomi-Sakurai
           Martin   crotylation with
           Periodinane   BF$_3$•OEt$_2$ or TiCl$_4$ O or undesired isomer

The attack of the nucleophilic face of the crotyl anion onto the face of aldehyde **N** is governed by which a Lewis acid catalyst is used, as shown in the next figure. TiCl$_4$ induces chelation-controlled addition, forming an undesired isomer. BF$_3$ induces Felkin–Anh attack, with the formation of the desired isomer **0**.

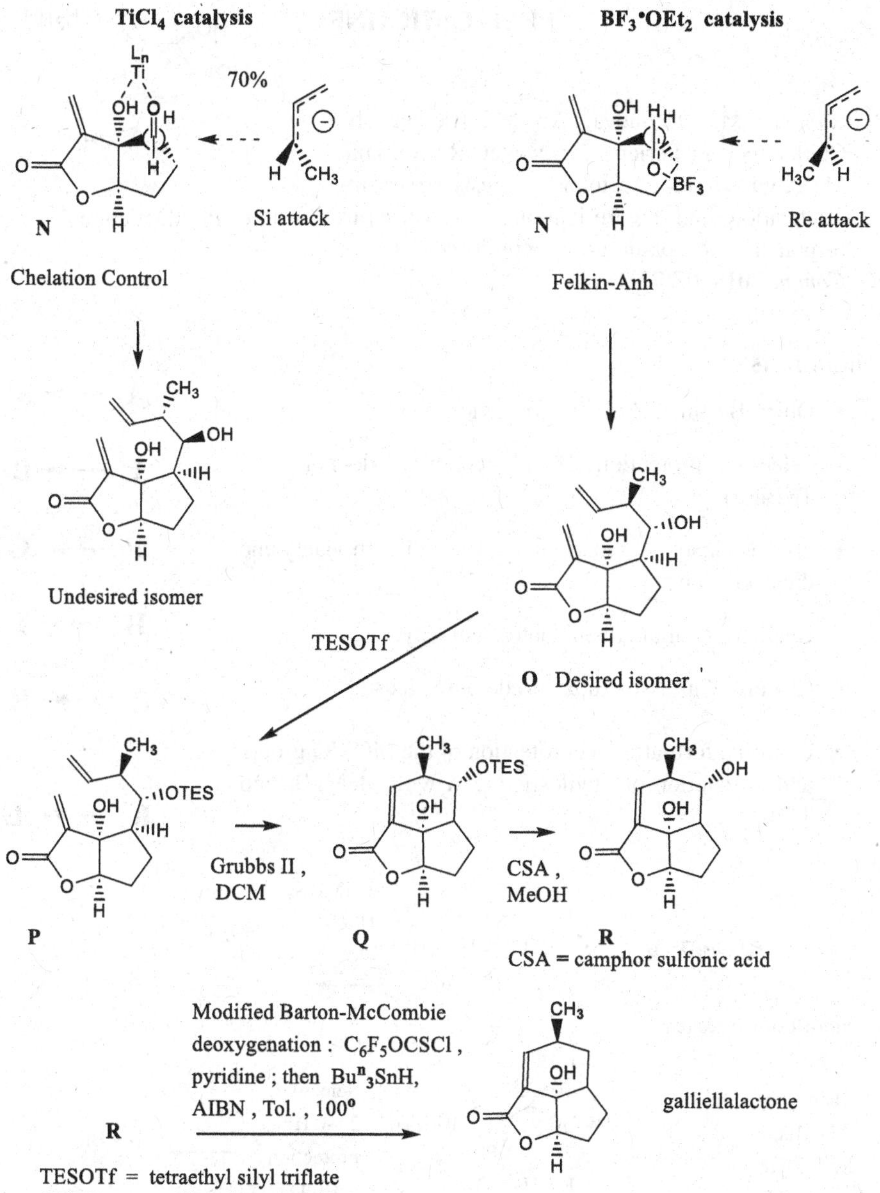

TiCl$_4$ catalysis   **BF$_3$·OEt$_2$ catalysis**

70%

N

Si attack

Chelation Control

N

Re attack

Felkin-Anh

Undesired isomer

TESOTf

O Desired isomer

P   Grubbs II ,
                        DCM

Q   CSA ,
                        MeOH

R

CSA = camphor sulfonic acid

Modified Barton-McCombie
deoxygenation : C$_6$F$_5$OCSCl ,
pyridine ; then Bu$^n_3$SnH,
AIBN , Tol. , 100°

R ⟶ galliellalactone

TESOTf = tetraethyl silyl triflate
AIBN   = azobisisobutyronitrile

PHYLLOSTICTINE

Riemer, M.; Uzunova, V. V.; Riemer, N.; Clarkson, G. J.; Periera, N.; Napier, R.; Shipman M. Phyllostictine A: total synthesis, structural verification and determination of substructure responsible for plant growth inhibition. *Chem. Comm.* **2018**, *67*, 7211–7214

L Phyllostictine A

HIGHLIGHTS

- Ohira-Bestman Z-alkene formation

  A ⟶ B

- Selective protection of a secondary alcohol with TriBnOT*

  C ⟶ D

- Oxy–Michael reaction on dimethtylactylene dicarboxyllate

  F ⟶ G

- Cyclization of an alkene onto an aldehyde

  H ⟶ I

- Closure of an N-methoxyamide onto an ester

  J ⟶ K

- Enamine formation, deprotection of the MOMO group, and production of Phyllostictine A with MeMgBr and TFA

  K ⟶ L

MeO_2C-$\overset{O}{\underset{}{P}}(OCH_2CF_3)_2$ **+** OHC-$(CH_2)_7$-H

A
Ohira-Bestman reagent

KHMDS
18-C-6
-78°

MeO_2C-$(CH_2)_6H$

B

Sharpless ADH :
AD Mix α ,
$MeSO_2NH_2$,
Bu^tOH/H_2O

$P = H$ **C**
(2S,3S)

Selective
2° -OH
protection
TriBnOT *

$P = Bn$

D

* TriBnOT = 2,4,6-tris(benzyloxy)-1,3,5-triazine =

1 MOMCl
2 DIBAL-H

D ⟶ **E**

Methoxymethyl
chloride
Dibutyl aluminum
hydride

Structure E: MOMO, HO, $(CH_2)_6H$, OBn

1. TIPS-OTf
2,6- Lutidine
2. H_2 Pd/C

⟶

Structure F: MOMO, TIPSO, $(CH_2)_6H$, **F** OH

MeO_2C—≡≡—CO_2Me

F ⟶

Acetylenic Oxy-Michael

Structure G: MOMO, TIPSO, $(CH_2)_6H$, MeO_2C O CO_2Me

G

HF•Py , PYR , THF ,
DCM -40° ;
TPAP , NMO , DCM ,
4 A MS

G ⟶

TPAP = tetrapropylammonium
          perruthenate

Structure H: MOMO, OHC, $(CH_2)_6H$, O, MeO_2C CO_2Me

H

NaHMDS ,
THF -78°

⟶

Structure I: MOMO, $(CH_2)_6H$, HOI·, O, MeO_2C CO_2Me

I 37%

Structure I: MOMO, $(CH_2)_6H$, HOI·, O, MeO_2C CO_2Me

I 37%

1M NaOH , MeOH ;
$MeONH_2$•HCl ,
WSD •HCl , TEA ,
HOBt•H_2O

⟶

Selective hydrolysis
and activation

WSD = water soluble
carbodiimide

Structure J: MOMO, $(CH_2)_6H$, HOI·, O, MeO_2C, =O, NHOMe

J 32%

TEA , DMF
60°

⟶

Structure K: MOMO, $(CH_2)_6H$, HOI·, O, O, N, O, Me

K 73%

MeMgBr , Et_2O ,
-78° ;
10% TFA

⟶

Structure L: HO, $(CH_2)_6H$, HOI·, O, O, N, CH_2, O, Me

**L Phillostictine A
35%**

KARRIKIN

"A natural product that has been found to signal plant growth."

Lachia, L.; Fonne-Pfister, R.; Screpanti, C.; StefanoRendane, C.S.; Renold, P.; Witmer, D.; Lumbroso, A.; Godineau, E.; Hueber, D.; De Mesmaeker, A. New and scalable access to Karrikin (KARI) and evaluation of its potential application on corn germination. *Helvetica Chem. ACTA* **2018**, *JOI*, e201800081

The synthesis of the rarely documented 4-H-pyran-4-ene substructure of Kerriken is an ingenious highlight of this report. The γ-pyrone derivative has been termed "pseudiaronatic."

The sequence of a Sonagashira reaction (forming **E**), followed by treatment with Brederick's reagent (forming **F**, with an enamine α to an oxygen), sets up the functionalities for a gold-catalysed intramolecular anti-Markovnikov Meyer–Schuster-type closure of intermediate **H** to the annulated rings of Kerrikin.

MERSICARPINE

"A highly oxidized indole alkaloid
with an atypical tetracyclic structure."

Yokoshima, S. Synthesis of natural products with polycyclic systems. *Chem. & Pharm. Bull.* **2013**, *61*, 251–257

MacAlpine, G.A.; Warkentin, J. Thermolysis of Δ ^{3}-1,3,4-oxadiazolin-2 ones and 2-phenylimino-Δ ^{3}-1,3,4-oxadiazolines derived from α,β-epoxyketones. An alternative method for the conversion of α,β-epoxyketones and alkynones and alkynols. *Can.J Chem.* **1978**, *56*, 308–315

Mersicarpine

A

- Two adjacent chiral tetrasubstituted carbons
- Enantioselective coupling α to a chiral imine auxilliary (**C** to **D**)
- Oxidative rearrangement of an epoxide adjacent to a semcarba-zide (**G**) into an alkyne aldehyde (**H**) through the Eschinmoser–Tanabe fragmentation
- Amine placement by attachment of a diazonium salt to C-3 of the indole (**J** to **K** to **M**)
- Fortuitous enantiospecific autooxidation – displacement at an indole (**M** to **A**)

B → Ph / NH$_2$ → **C** → ethyl acrylate, hydroquinone, AcOH, aq. THF → **D** → IBX , DMSO

E → H$_2$O$_2$, NaOH , H$_2$O , MeOH → **F** → H$_2$N—CO—NHNH$_2$ → **G**

G → Eschinmoser-Tanabe fragmentation / Pb(AcO)$_4$, DCM "oxidation and thermolysis" → 1,3,4,oxadiazoline X = O , NR Warkentin's intermediate →

H 60 % → NaBH$_4$; o-Iodoaniline Pd(Ph$_3$P)$_4$, CuI , DMF , TEA , 80° → **I** → Helmitsberger-Kittle reaction NaAuCl$_4$•2H$_2$O EtOH →

J → Benzene diazonium chloride , NaOAc , isopropanol , dioxane → **K** 97 %

K → NaH , Tol ; MsCl , TEA → **L** 77%

L → H$_2$, Pd/C , Pr^{i}OH, DCM → **M** → NaHCO$_3$, Pr^{i}OH, DCM ; autooxidation Me$_2$S → Mersicarpine **A**

LYCOPOSERRAMINE

Ishida, H.; Kimura, S.; Kogure, N.; Kitajima, M.; Takayama, H. Total synthesis of (+/-)-lycoserramine-R, a novel skeletal type of Lycopodium alkaloid. *Tet.* **2015**, *71*, 51–56

Lycoposerramine
14

Two dimensional
rotated image

Hajos Parrish ketone

Hajos-Parrish inspired
Key Intermediate **X**

As the pseudo three-dimensional image of lycoserramine above is flattened and rotated, the possibilities of a Hajos–Parrish-type starting material can be seen for conversion into the hexahydroindene core structure of lyciposerramine. The key intermediate **X** was then designed to annulate the two nitrogen-containing rings with that core structure at the carbonyl groups. A Michael reaction would hopefully deliver a methyl group against the α, β unsaturated carbonyl group of X in the needed orientation. The following retrosynthetic plan was guided by i) previous syntheses of the Hajos–Parrish ketone, ii) the Wharton rearrangement, for reversal of the position of the α, β unsaturated carbonyl, and iii) the several known schemes for the annulation of piperidine, pyridine, and pyridinone motifs onto cyclic ketones.

RETROSYNTHETIC PLAN

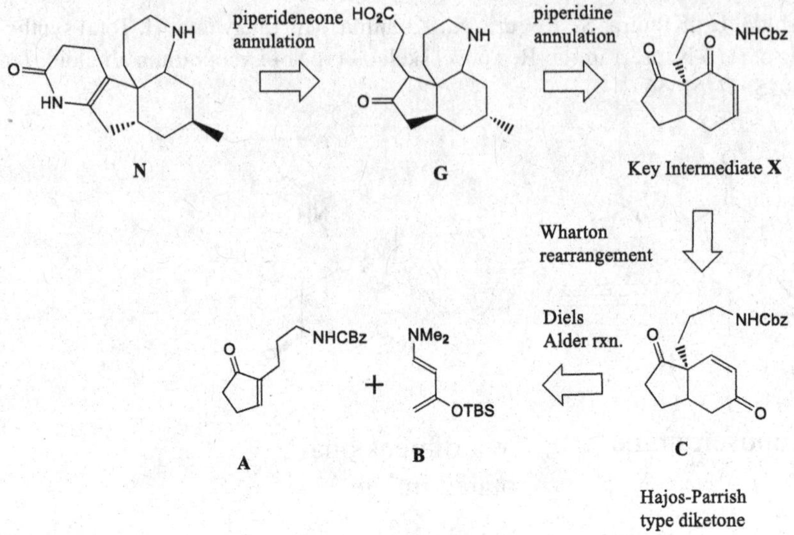

N   **G**   Key Intermediate **X**

Wharton
rearrangement

Diels
Alder rxn.

A   +   **B**   **C**

Hajos-Parrish
type diketone

In the final steps of the synthesis, a methyl group would be
appended to the α, β unsaturated carbonyl group of **X**, and two
piperidine groups to the cyclic ketones. A Wharton rearrangement was planned to realize a 1,3 carbonyl shift to transform
the Hajos–Parrish type diketone **C** into Key Intermediate X.A
Diels-Alder approach was to be used in the preparation of **C**.

EXPERIMENTAL SEQUENCE

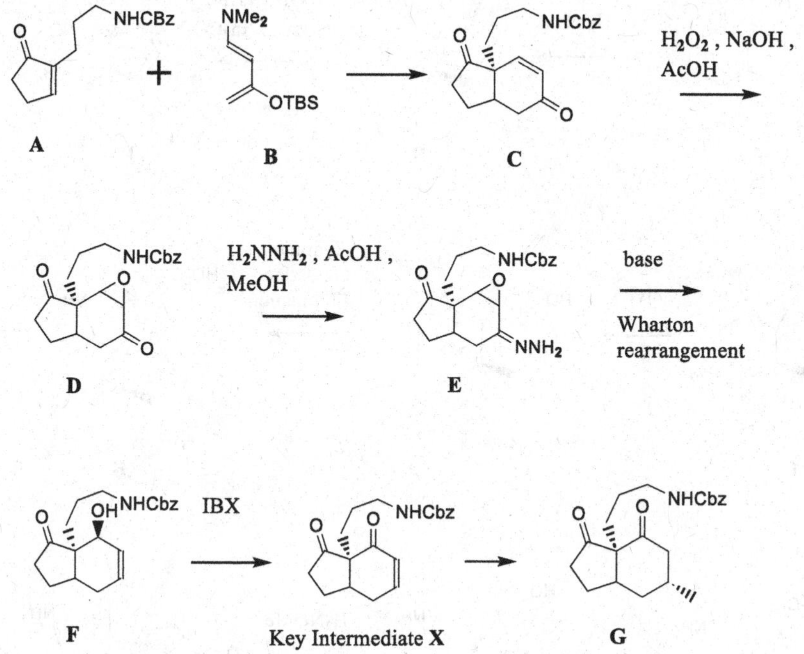

A + B → C

H$_2$O$_2$, NaOH, AcOH →

D

H$_2$NNH$_2$, AcOH, MeOH →

E

base →

Wharton rearrangement

F

IBX →

Key Intermediate X

→ G

Lycoposerramine

4 Diversification and Simplification

The modification of organic compounds for biomedical or industrial applications is done to improve their properties, identify the essential structural features, eliminate unfavorable features, or prepare equivalent derivatives that are better suited for preparation and use.

During a total synthesis, less complex versions of a desired compound may be encountered. Such versions are sometimes useful in discovering, for example, the moieties responsible for adhesion, binding to an enzyme, causing an unwanted side effect. The simplified modifications described in this chapter – of abyssomycin, epothilone, conolidine, vancomycin, and sphingosine – are examples of this approach.

Other examples are syntheses designed to facilitate late stage derivatization, for example, with altered lipophilicity for penetration of the blood brain barrier, plasma stability and specificity towards certain organs, and enzyme receptors are alternate objectives. The examples of Fuligonaden, Kaurene, GABA drugs, and diazepine are described here.

DOI: 10.1201/9781003397816-4

TWO AMINOPHILIC REAGENTS

I. Onchidal and Derivative

Cadelis, N.M.; Copp, B.R. Investigation of the electrophilic reactivity of the biologically active marine sesquiterpenoid onchidal and model compounds *Beilstein J. of Org. Chem.* **2018**, *14*, 2229–2235

S. Li; X. Wang; L. Liu; J. Kang; L. Wang; H. Liu; C. Ruan; A. Nie; Z. Zheng; Y. Xie; G. Zhao; J. Xiao; Y. Hu; W. Zhong; H. Cui; and X. Zhou. Chinese Patent CN Pat. PCT/CN2004/001118, April 6, 2006

5-membered-S-heterocyclic compounds and their use in preparing of medicines for treating or preventing the obesity-relating Diseases.

Onchidal is the defensive secretion of a Pacific mollusk. It has found use in attaching theranostic groups to proteins. The 1-acetoxy-3-formyl-1,3-butadiene moiety attaches to two primary amines, as illustrated in the model reaction below.

A speculative mechanism involves the initial formation of an intermediate substituted itaconic dialdehyde, which condenses with the first amine equivalent to form the electrophilic iminium salt S. In a Michael-type addition, the second amine yields pyrrolidine diamine **A**.

SYNTHESIS OF ONCHIDAL ANALOG E

The Chinese patent describes the synthesis of the Homer–Wittig reagent **B**, used here for the preparation of itaconic ester derivatives **C**. Reduction and oxidation of **C** to dialdehyde **D** is followed by selective acetylation of the terminal aldehyde, forming the l-acetoxy-3-formyl-1,3-butadiene moiety in **E** below.

Chinese Patent CN $\longrightarrow$

EtO, P, EtO, O

CO_2Et
CO_2Et

B

$H(CH_2)_5CHO$ $\longrightarrow$

$H(CH_2)_5$ CO_2Et CO_2Et

C both E and Z

C $\xrightarrow{\text{LAlH}_4;\ \text{Des Martin [O]}}$

$H(CH_2)_5$ CHO CHO

D (E/Z)

$\xrightarrow{\text{Ac}_2\text{O}}$

$H(CH_2)_5$ OAc CHO

E (E/Z)

The availability of such Diels–Alder dienes as **E** make them potential starting materials for the preparation of 1,3,4-trisubstituted-2- cyclohexenes.

II. COLORIMETRIC AMINOPHILE

Diaz, Y.J.; Zachariah, A.; Page, Z.A.; Knight, A.S.; Treat, N.J.; Hemmer, J.R.; Hawker, C.J.; de Alaniz, J.R. A versatile and highly selective colorimetric sensor for the detection of amines. *Chem. Europ. J* **2017**, *23*, 3562–3566

MAF $+ R^1R^2NH \longrightarrow R^1R^2N$ Discrimination at 532 nm. ; 1° Amine pink : 2° Amine blue

"The utility of Meldrum's activated furan (**MAF**) for the colorimetric detetection of sub-ppm levels of amines in solution, on solid supports and as vapors, is reported. The utility of this novel system in thin layer chromatography, solid-phase peptide synthesis, and food spoilage detection is demonstrated."

There is a curious analogy here with the acidic dinitrophenylhydrazine reagent that yields red, orange, or yellow hydrazone precipitates depending on the aliphatic or aromatic substituents of the aldehydes or ketones with which it reacts.

TRUNCATED ABYSSOMICIN

Monjas, L.; Fodran, P.; Kollback, J.; Cassani, C.; Olsson, T.; Genheden, M.; Joakim Larsson, D.G.; Wallentin, C.-J. Synthesis and biological evaluation of truncated derivatives of abyssomicin as bacterial agents. *Beilstein J. Org. Chem.* **2019**, *15*, 1468–1474

Abyssomicin is a natural product of potential therapeutic value. Its complex structure precludes near term synthesis of enough material for enzymatic binding or broad-spectrum antibiotic screening. In a preliminary probe into a synthesis, a truncated analog was designed, to be assembled from the loops **C** and **G** as shown below.

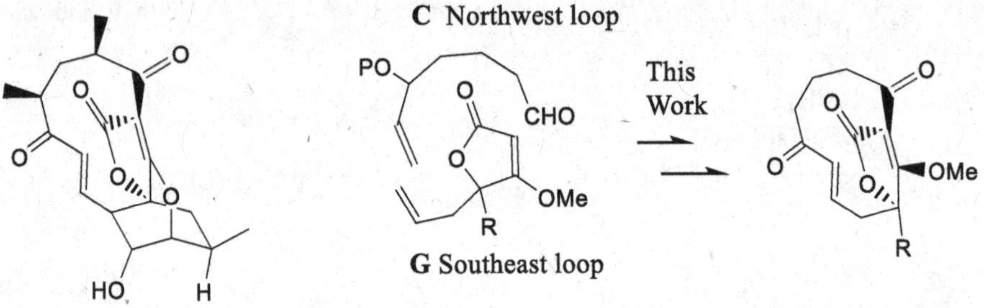

C Northwest loop

This Work

G Southeast loop

Abyssomicin

Truncated Analog I

Synthesis of Northwest Loop C

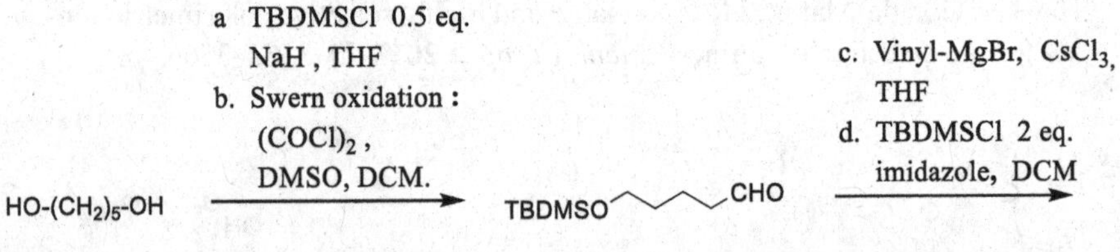

a TBDMSCl 0.5 eq.
   NaH , THF
b. Swern oxidation :
   (COCl)$_2$,
   DMSO, DCM.

HO-(CH$_2$)$_5$-OH

TBDMS = *tert*-butyl dimethylsilyl

TBDMSO⌒⌒⌒CHO

A

c. Vinyl-MgBr, CsCl$_3$, THF
d. TBDMSCl 2 eq. imidazole, DCM

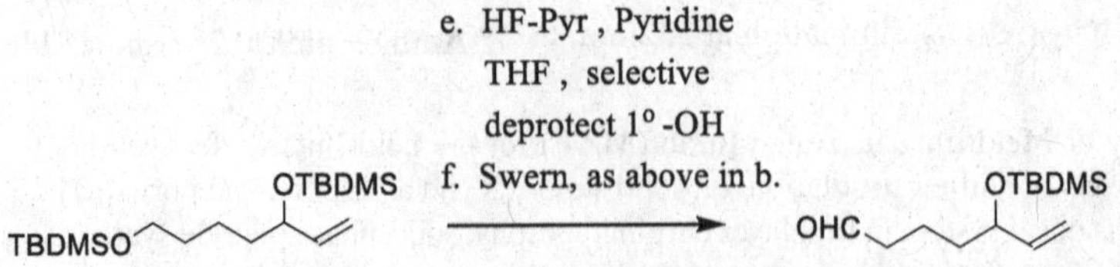

e. HF-Pyr , Pyridine
   THF , selective
   deprotect 1°-OH
f. Swern, as above in b.

OTBDMS

TBDMSO⌒⌒⌒

B

OHC⌒⌒

OTBDMS

C   Northwest loop

SYNTHESIS OF SOUTHEAST LOOP G

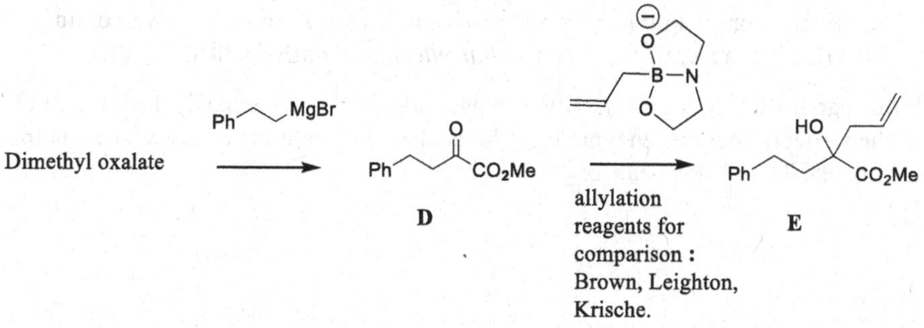

Dimethyl oxalate → (Ph ∕∕ MgBr) → D (Ph CH₂CH₂ C(O) CO₂Me)

allylation
reagents for
comparison :
Brown, Leighton,
Krische.

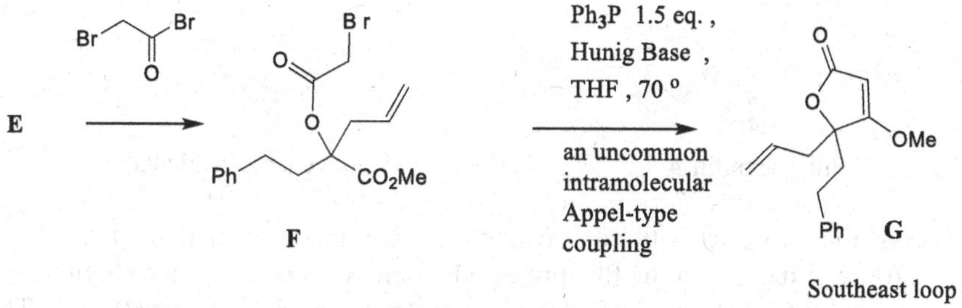

E

E → F (Br CH₂ C(O) O-C(CH₂CH₂Ph)(CO₂Me)(allyl))

Ph₃P 1.5 eq. ,
Hunig Base ,
THF , 70 °

an uncommon
intramolecular
Appel-type
coupling

G

Southeast loop

CLOSING THE LOOPS

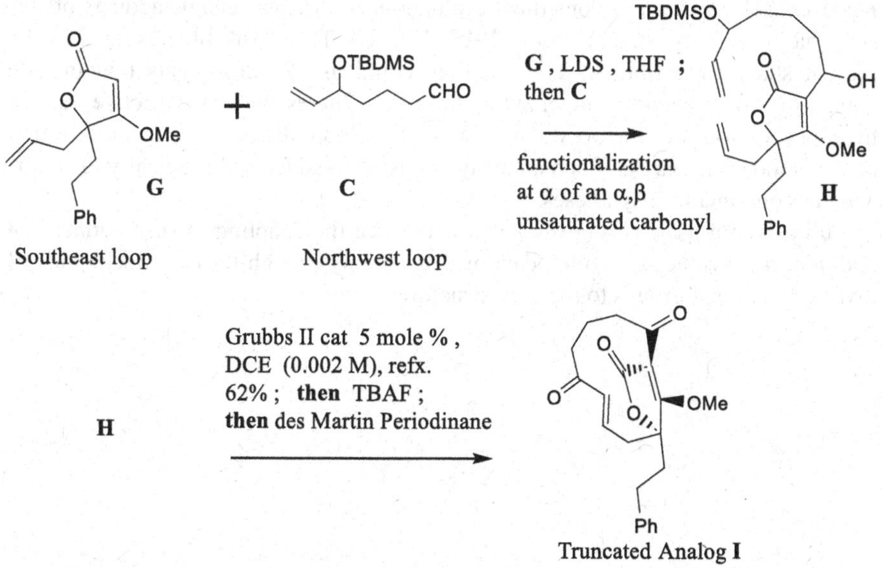

G
Southeast loop

+

C
Northwest loop

G , LDS , THF ;
then C

functionalization
at α of an α,β
unsaturated carbonyl

H

Grubbs II cat 5 mole % ,
DCE (0.002 M), refx.
62% ; then TBAF ;
then des Martin Periodinane

H

Truncated Analog I

KNOEVENAGEL ADDUCTS OF FULIGOCANDIN B

Arai, M.A.; Masuda, A.; Suganami, A.; Tamura, Y.; Ishibashi, M.; Synthesis and Evaluation of Fuligocandin B Derivatives with Activity for Overcoming TRAIL Resistance. *Chemical and Pharmaceutical Bulletin* **2018**, *66*, 810–817

Fuligocandin B itself, and its Knoevenagel adducts, were prepared for evaluation of their effectiveness as enzyme ligands involved in tumor necrosis factor-inducing effects like TRAIL Resistance.

Fuligocandin A   Knoevenagle Adducts

The synthesis begins with consecutive selective amide formations, the second being closure of the ring at the Boc protected amine (**C** to **D**). Selective alkynylation is then realized at the more electrophilic Boc protected amide carbonyl (**D** to **E**). The subsequent Meyer–Schuster-like rearrangement (**E** to **H**) delivers the natural product in 99% yield.

The terminal methyl ketone then allows for the facile formation of a library of Knoevenagel adducts for biomedical evaluation. Such Knoevenagle adducts of steroids have been reported (*Steroids* **2015**, *101*, 47). The R substituents used in the present study were limited. The coupling could use R substituents that include competitive or nonconpetitive enzyme inhibitors, fluorescent or radioactive tracers, hydrophilic groups to impart water solubility, or lipophilic moieties for penetration of the blood brain barrier. Such an array of desirable adducts are typically ideal targets for combinatorial syntheses.

Fuligocandin itself is robust enough to survive the coupling to such adducts. In addition, this synthesis of fuligocandin B is flexible enough for the attachments of diverse functional groups to the core structure.

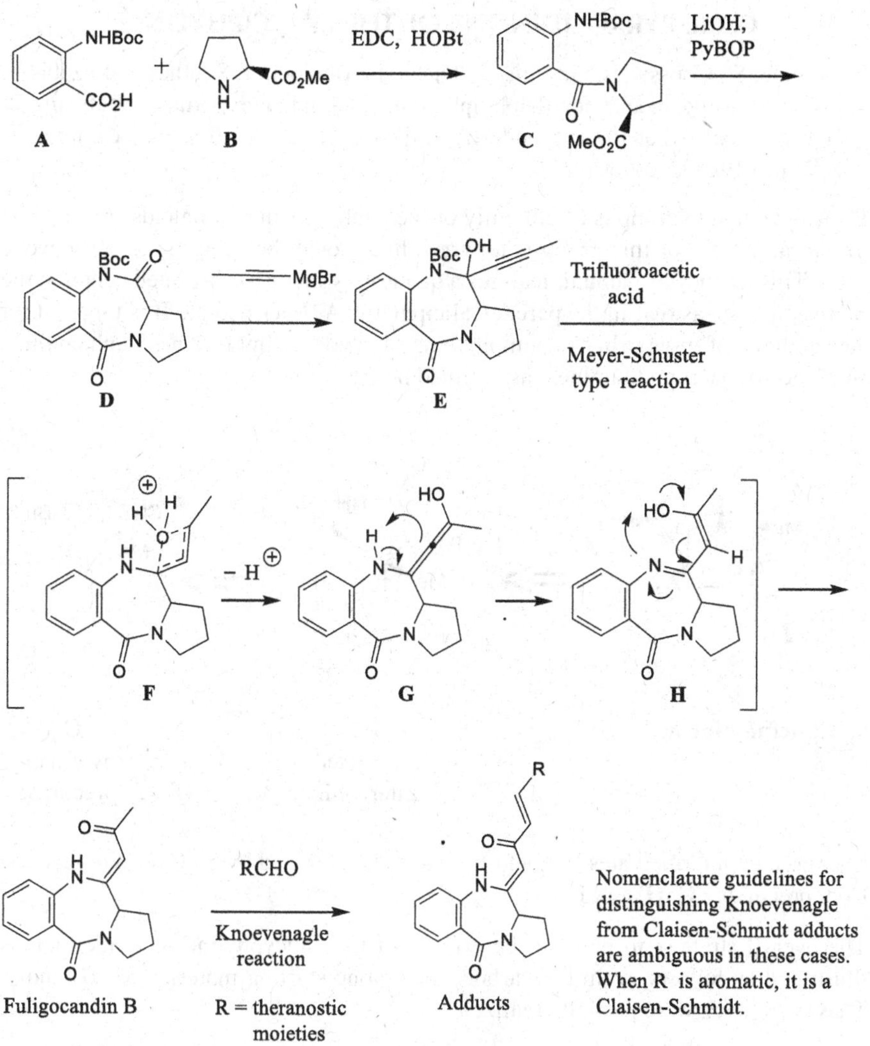

Fuligocandin B

R = theranostic moieties

Adducts

Nomenclature guidelines for distinguishing Knoevenagle from Claisen-Schmidt adducts are ambiguous in these cases. When R is aromatic, it is a Claisen-Schmidt.

CORE PYRROLIDINONE OF THE CALYCIPHYLINES

Jansana, S.; Coussanes, G.; Puig, J.; Diaba, F.; Bonjoch, J. Synthesis of azabi-cyclic building blocks for delphniphyllum alkaloid intermediates featuring N-trichloroacetyl enamide 5-endo-trig radical cyclizations. *Helvetica Chimica ACTA* **2019**, *102*, e1900188

The calyciphyline group is a subfamily of the daphniphyllum alkaloids. A common structural feature of this group is a pyrrolidine moiety bridging three carbocyclic rings. This azabicyclic subunit features a quaternary methyl-substituted center at the cis fusion, such as that in the parent Caliciphyline A. Intermediate **B** is targeted for the synthesis of three **other** daphniphyllum alkaloids - Himalensine, Daphenylline, and 2-deoxymacropoddumine – as outlined below.

| **Caliciphyline A** | **B** | **C** |
| | Targeted intermediate | Overman precursor |

The targeted intermediates F, I, and **L** were to be obtained by an Overman reaction from precursors **C**, **H**, and **K**.

The overall strategy to prepare these three of the calyciphyline-type alkaloids is illustrated as follows, from the carbocyclic ketone starting materials **D**, **G**, and **J**. These syntheses were partially realized.

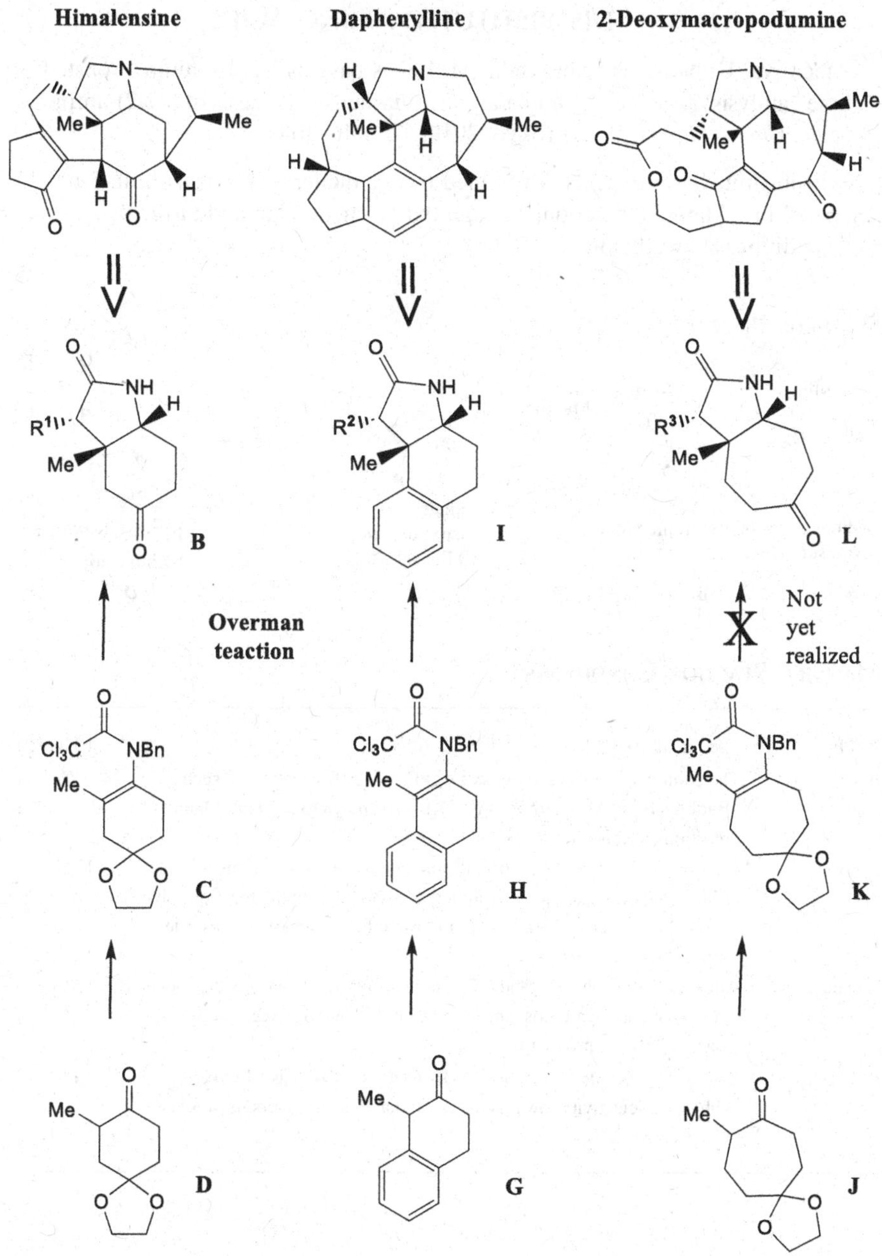
Himalensine
Daphenylline
2-Deoxymacropodumine
B
I
L
Overman
teaction
Not
yet
realized
C
H
K
D
G
J

B ^{2,2}-DISUBSTITUTED AMINO ACID

Amemiya, F.; Noda, H.; Shibasaki, M. Lewis base assisted lithium Brønsted base catalysis: A new entry for catalytic asymmetric synthesis of β(2,2)-amino acids. *Chem. Pharm. Bull.* (Tokyo) **2019**, *67*, 1046–1049

β ^{2,2}-disubstituted amino acids are formed from a quaternary isoxazolidin-5-one **C**, generated in a Mannich reaction, through the synthetic plan and under the customized conditions shown below.

SYNTHETIC PLAN

		quaternary	
Aldimine	Isoxazolidin-5-one	isoxazolidin-5-one	β^{2,2}-disubstituted
precursor		Michael adduct	amino acid
A	**B**	**C**	**D**

MANNICH REACTION CONDITIONS

AgPFO$_6$	Used in all trials	10 mole%
Lewis Acid	The combination of the silver salt AgPF$_6$ and a Brønsted base, such as Barton's base (Me$_2$(C=NBu^t)-NMe$_2$), was known to promote Mannich reactions of this type.	10%
Brønsted Base	A strong base is needed to form the enolate of the isoxazolidin-5-one. This is the origin of epimerization that occurs during the reaction. Five strong bases were screened. Lithium **p**-methoxy phenoxide was selected for evaluation of the ligands.	12 %
Ligand	Among the phosphine ligands, Taniaphos consistently gave better yields, favorable anti /syn ratios, and a greater enantiomeric excess of the desired anti 4-S product **C**.	12 %
Lewis Base	Bisphosphineoxide ligands Ph(P=O)-(CH$_2$)$_n$-(P=O)Ph as additives enhanced selectivity towards the desired anti enantiomers most when n = 4 was used	10 %

THE MANNICH ADDUCTS

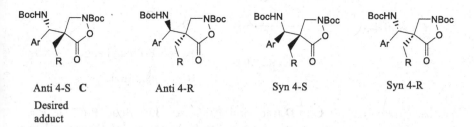

A + B

Mannich reaction conditions
in table below

Diasteriomers of the quaternary isoxazolidin-5-ones **C**

Anti 4-S **C**　　　Anti 4-R　　　Syn 4-S　　　Syn 4-R
Desired
adduct

Conditions were optimized for yield and selectivity towards the desired Anti 4-S product **C**.

Ag Salt	Ligand	Lewis Base	Brønsted Base	% Yield	Anti/Syn	er (anti)
AgPF$_6$	(R) Segphos	Barton's Base	–	90	51/49	53/47
AgPF6	(R, R$_p$) Taniaphos	DBU	–	68	60/40	70/30
AgPF$_6$	(R, R$_p$) Taniaphos	Phenoxide	Bisphosphine	72	61/39	91/9 *optimal*

LiO—⟨⟩—OMe

Phenoxide

Bisphosphine Oxide

Barton's Base

(R, R$_P$) Taniaphos　　　　**(R) Segphos**

Beta Amino Acid Formation

Deprotection and ring opening of the Anti 4-S **C** adduct with trifluoro acetic acid, followed by treatment with phenylpyruvic acid **E** "triggered a decarboxylative amide formation with the hydroxylamine moiety in a chemoselective fashion" with the formation of **D**. The transformation gives rise to a Hunsdiecker or Minisci type product.

Ar = 4-F-Phenyl and 3-F-Phenyl

R = 2-naphthyl

For background on this **C** to **D** transformmation, see: J.W. Bode, Prof. Dr. R.M. Fox and K.D. Baucom. Chemoselective Amide Ligations by Decarboxylative Condensations of N-Alkylhydroxylamines and α-Ketoacids. *Angewandte Chemie* 2006, **45**, *8*, 1248

CONFORMATIONALLY CONSTRAINED EPOTHILONE

Kuzniewski, C.N.; Glauser,S.; Gaugaz, F.Z.; Schiess, R.; Rodriguez-Salarichs, J.; Vetterli, S.; Horlacher, O.P.; Gertsch, J.; Redondo-Horcajo, M.; Canales, A.; Jimenez-Barbero, J.; Diaz, J.F.; Altmann, K.-H. Synthesis, profiling, and bio-active conformation of trans-cyclopropyl epothilones. *Helvetica Chem. ACTA* **2019**, *102*, e1900078

Analogs of the epothilone family have been studied to discover one that retains the original activity and is available in a scalable synthesis. One of the projected candidates is 1-Bl.

Epothilone B   1-B1 Projected candidate

A specific objective of the work was to replace the epoxide moiety with a more synthetically accessable functionality.

D
Northeast corner of 1-B1
This synthesis.

Substitution of the thiazole group with the benzothiazole group was found to partially compensate, in metabolic effect, for the absence of the southern hydroxyl group. The cyclopropyl group restricts the conformational degrees of freedom into a more bioactive configuration.

The synthesis features an enantioselective Brown allylation using isopinocamphenyl borane, and Andre Charette's enantioselective cyclopropanation, using an unconventional boron-based chiral catalyst.

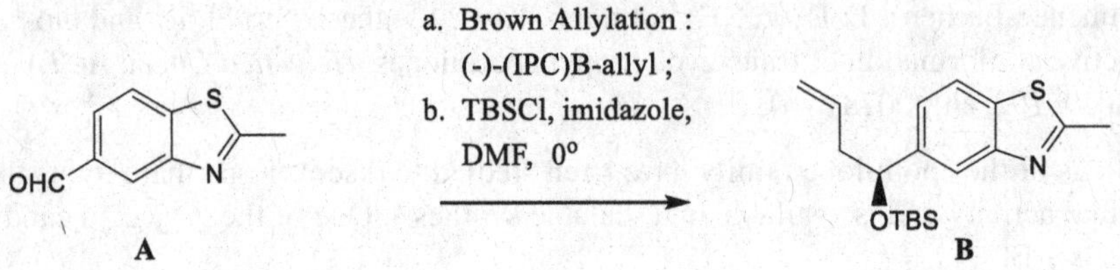
OHC
S
N
A
a. Brown Allylation :
(-)-(IPC)B-allyl ;
b. TBSCl, imidazole,
DMF, 0°
OTBS
B

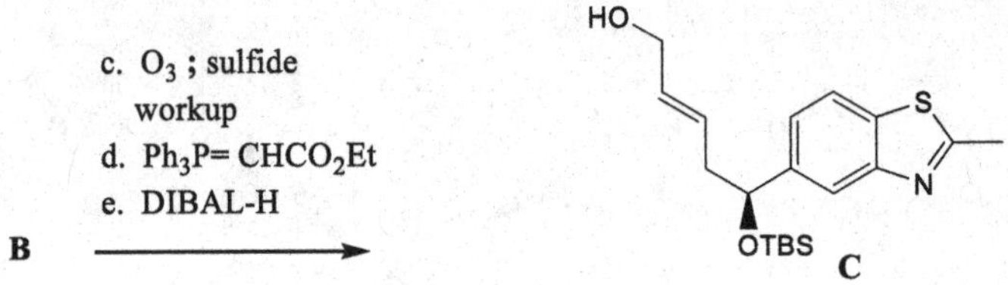
B
c. O3 ; sulfide
workup
d. Ph3P= CHCO2Et
e. DIBAL-H
HO
S
N
OTBS
C

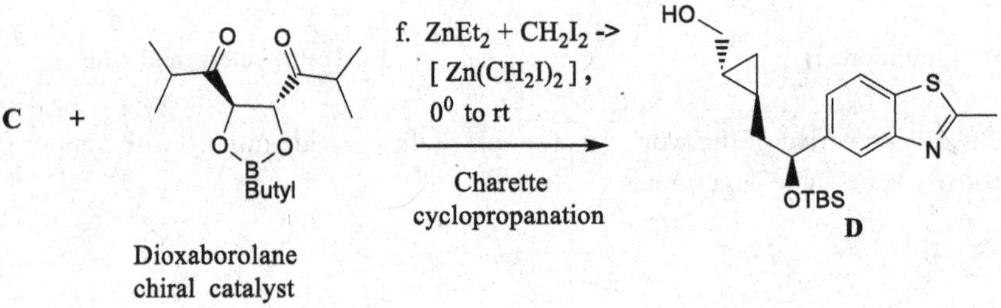
C +
O
O
B
Butyl
Dioxaborolane
chiral catalyst
f. ZnEt2 + CH2I2 ->
[Zn(CH2I)2] ,
0° to rt
Charette
cyclopropanation
HO
S
N
OTBS
D

The synthesis of 1-Bl from D was realized using previously published methods, including metathesis and macrolactonization.

KAURENE DERIVATIVE

Hu, Y.; Li, X-N.; Ma, Z-J.; Puno, P.T.; Zhao, Y.; Zhao, Y.; Xiao, Ye-Zhi.; Zhang, W.; Liu, J. Synthesis of Novel ent-Kaurane-Type Diterpenoid Derivatives Effective for Highly Agressive Tumor Cells. *Molecules (Basel, Switzerland)* **2018**, *23*, 2727

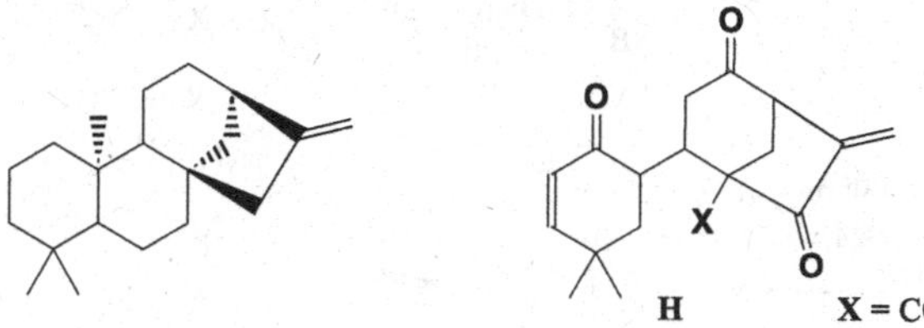

ent-kar-16-ene   bioactive kaurene derivative

Kaurene is a biosynthetic precursor of the gibberellins involved in plant growth. In a related synthesis project, some of the intermediates were screened for anticancer activity. Intermediate **H** showed growth inhibition in certain cell lines. It had been prepared in a Michael coupling between **A** and **D** illustrated in the highly simplified retrosynthetic plan shown below.

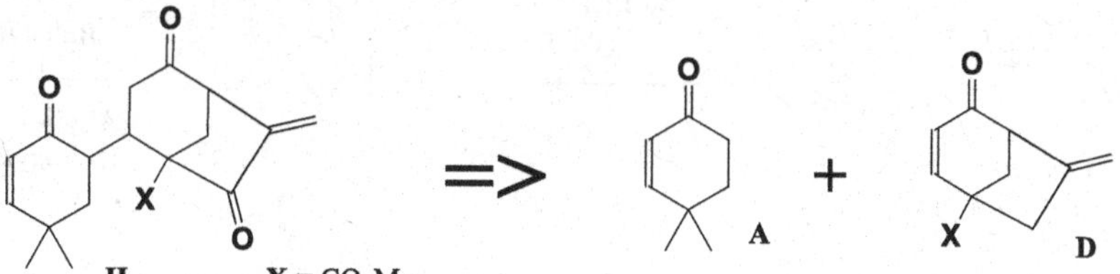

H   X = CO$_2$Me

kaurene derivative

A Robinson reaction delivered **A** in a single step.

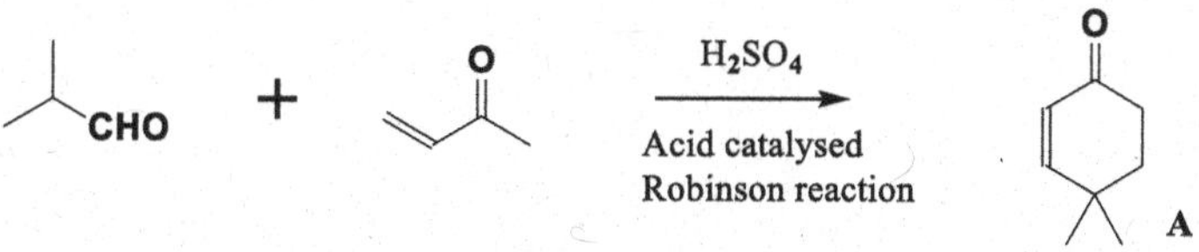

The synthesis of component **D** involved two successive additions at the carbon α to
the carbonyl of intermediate **B**.

The Michael reaction of **A** with **D** was followed by the hydroxylation and oxidation
at the allyllic carbon.

CONOLIDINE SIMPLIFIED

Arita, T.; Asano, M.; Kubota, K.; Dornon, Y.; Machinaga, N.; Shimada, K. Discovery of conolidine derivative DS39201083 as a potent novel analgesic without μ opioid activity. *Bioorg. Med. Chem. Lett.* **2019**, 29, 1938–1942

Naoe, S.; Yoshida, Y.; Oishi, S.; Fujii, N.; Ohno, H. Total synthesis of (+)-conolidine by the gold(I)-catalysed cascade cyclization of a conjugated enyne. *J Org. Chem.* **2016**, *81*, 5690–5698

Derivatives of conolidine are prepared and screened in search of any with the desired anesthetic effects, but without μ activity, which has been associated with itching, addiction, constipation, respiration and depression. Certain of these targeted derivatives expressed the desired high analgesic activity and low addictive effects of conolidine. Their ease of preparation make them promising candidates for further study and scale-up.

Conolidine   Targeted Derivitives

Among the many derivatives screened, the R^3 = methyl, R^1 = R^2 = H candidate designated "DS39201083" was selected as a hit for its comparatively concise synthesis, good analgesic profile, and a low euphoria score in a mouse addiction model.

The N-protected indole **D** is functionalized at C-2 (**D** to **E**), then detosylated and oxidized to the ketone **G**. This activated the indole at C-3 towards intramolecular Mannich cyclization in the next step. In alternate preparations of the targeted derivatives, substituted indoles and piperidines **C** were used for the syntheses of intermediate **E**.

a LiAlH$_4$ ⟶   X = CH$_2$OH   **B**

A X = CO$_2$H
R^3 = CH$_3$ and
others

b SO$_3$, Pyridine, DMSO,
Hunig Base, DCM   ⟶   X = CHO   **C**
(Parikh Doering [O])

c BuLi   ⟶   **E**

P = SO$_2$Ph

C   +   **D**

d Cs$_2$CO$_3$, MeOH
an alternative
detosylation agent
to Mg/MeOH   ⟶   P = H   **F**

e MnO2   ⟶   **G**

The derivatives incorporated various
combinations of the following substituents

R^1 = H , CH$_3$, CH$_2$CH$_2$OH
R^2 = H , OCH$_3$
R^3 = H , CH$_3$
R^4 = H , CH$_3$, Ethyl , sec-Butyl

G ⟶

f HCl
g paraformaldehyde

Targeted Derivitives

ANNULATED THIOPHENE

OHC and annulated thiophene structure with C_2H_5 and $(CH_2)_nPh$ on N

Baumeister, S.; Schepmann, D.; Wunsch, B. Thiophene bioisosteres of GluN2B selective NMDA receptor antagonists: Synthesis and pharmacological evaluation of [7]annulo[b]thiophen-6-amines. *Bioorg. Med. Chem.* **2020**, *28*, #115245

The targeted [2,3] annulated thiophene compounds were designed for the therapy of neurodegenerative disorders.

A

BuLi , THF , -78° ;
CO_2 , THF   77%

CO_2H
[2,3] thiophene
dicarboxyllic acid
B 77%. (see Addendum)

$LiAlH_4$;
PBr_3 , $0°$

C (Br, Br)

dimethyl-3-oxyglutarate ,
K_2CO_3 ,

CO_2Me ... Br CO_2Me ... O
D

CO_2Me ... O ... CO_2Me
E

HCl

F (thiophene, O)

d) $Ph(CH_2)_nNH_2$,
$NaB(OAc)_3$
n = 1 - 4

R
N
$(CH_2)_nPh$

G R = H

H R = CH_2CH_3

CH_3CHO ,
$NaBH(OAc)_3$

BuLi , THF ;
N-Formyl Piperidine

OHC and annulated thiophene with C_2H_5 and $(CH_2)_nPh$ on N

Targeted compounds

Optimal binding
properties realized
for n = 3 and 4

Among the numerous CO sources in current use, N-formylpiperidine was most convenient for the formylation of **H**: both regioisomers of the aldehyde were separable. This provided an extensible functionality for the attachment of other binding modifiers, immobilization onto a support, or attachment to a delivery vehicle.

ADDENDUM

In the course of this study it was found that thiophene 3-carboxyllic acid yields only traces of the 3,4 dicarboxyllic acid under the same conditions.

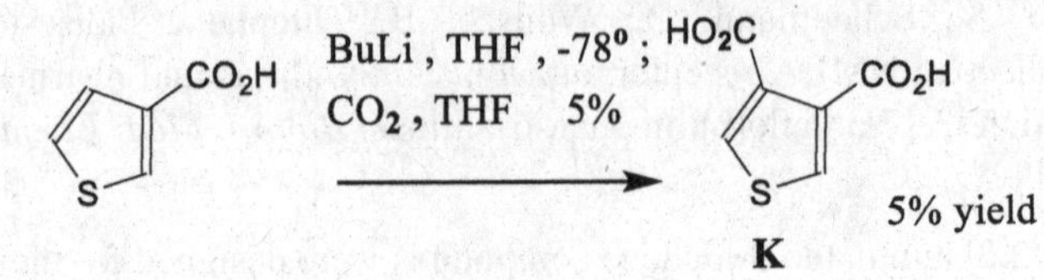

$$\text{K}$$

5% yield

This meant that the preparation of the analogous [3,4] annulated thiophene could not be realized by this route at this time. It should be noted that 3,4-diaminothiophene **J** is rarely available commercially, although it van be prepared by the reduction of 1,2-dibromo-2,3-dinitrothiophene *(Biorg. Med. Chem.* **2018**, *26*, 1628) as shown below. Conceptually feasible routes from this diamine **J** to the needed 3,4-thiophene dicarboxyllic acid **K** might be developed, through bis-diazotization, bis-amide formation, or a diisonitrile intermediate.

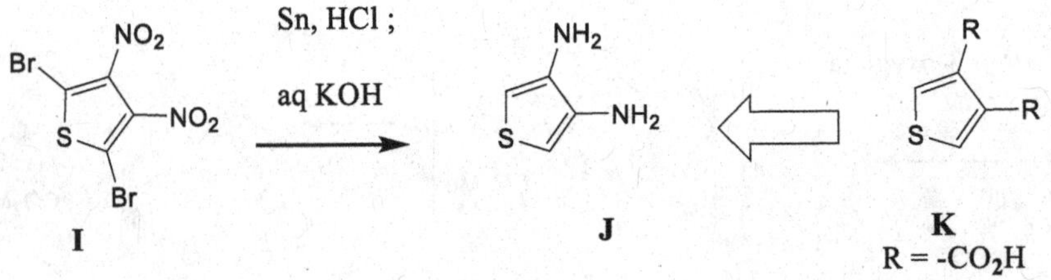

I   J   K
                                                          R = -CO₂H

MODIFIED PYRAZOLAQUINOLINONE DRUG

Iorio, M.T.; Rehman, S.; Mampali, K.; Stoeger, B. ... Mihovilovic, M.D. Variations on a scaffold – Novel GABA-A receptor modulators. *Eur. J. Med. Chem.* **2019**, *180*, 340–349

- A Pyrazolaquinolinone drug **PQ** was found to be highly active on α1β3 GABA receptors.
- A novel indole scaffold A was suggested by the pharmacologic descriptors of PQ.
- The indole core of the target scaffolds was prepared by a Fischer Indole synthesis.
- The hydrazine linkage was realized with a diazononium salt coupling.
- Differently substituted analogs of the target scaffold, at R^1, R^2, and R at the carboxylate functionality, were studied for their binding to the GABA receptors.

GABA receptor binding Pyrazoloquinoline

PQ

Indole Scaffold

A

Some GABA receptors have allosteric binding sites which are located on a different part of the receptor surface from the actual binding site. **PQ** is an allosteric modulator which acts in this way, by either enhancing the activity of the GABA receptor or diminishing it. As such, it could act as either an agonist or an antagonist, depanding on the nature of the substituents R^1, R^2, and R^3.

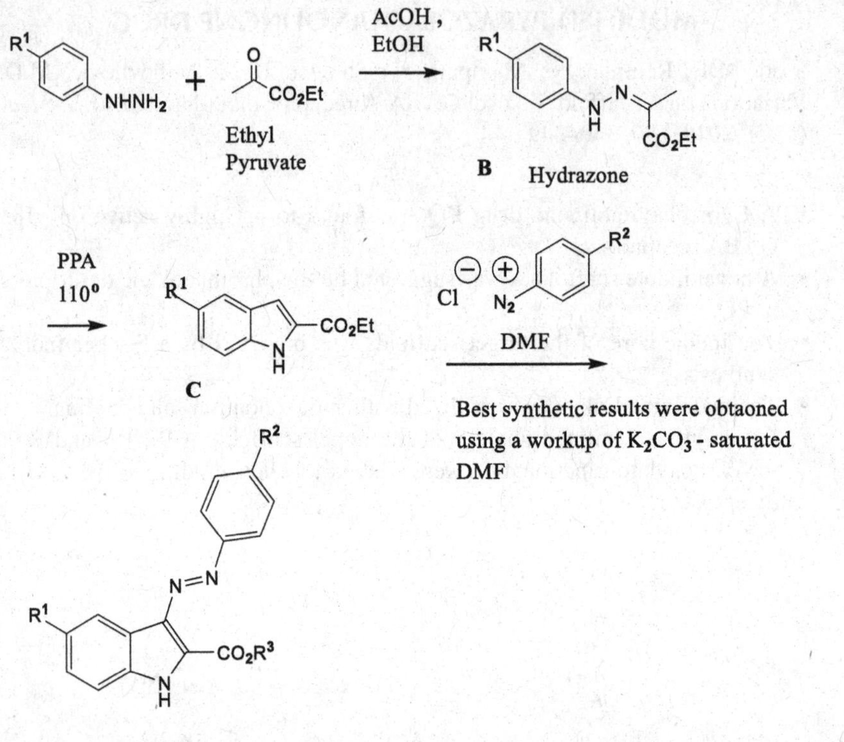

B   Hydrazone

C

A

Best binding results were obtained with

$R^1 = Br$, $R^2 = OMe$, $R^3 = H$

THE BISPIRO CORE OF MOLLANOL A

Miao, J.; Zheng, Y.-L.; Wang, L.; Lu, S.-C.; Zhang, S.P.; Gong, Y.-L.; Xu, S.
Towards the total synthesis of grayanane diterpene mollanol A by a Prins [3+2]
strategy. *Org. Biomol. Chem.* **2020**, *18*, 1877–1880

Precursor **G** is targetted as a key intermediate in the synthesis of Mollinol A.

Mollanol A   Precursor **G**

This work

B +   **C**   Prins
canditions   →   **D**   TMS   Oxybromination   →   **F**   Debromination

An acyclic coupling of alkene **B** with chiral aldehyde **C** under Lewis acid conditions yields the Prins product **D**. An insightful bromination- oxygen displacement
closed the middle bispiro ring, forming **F**.

THE SYNTHESIS

A bispiro structure with the new chiral center is assembled from the Prins product **D** in an oxybromination sequence (**D** to **E** to **F**) shown below. Free radical debromination forms Precursor **G**, as follows.

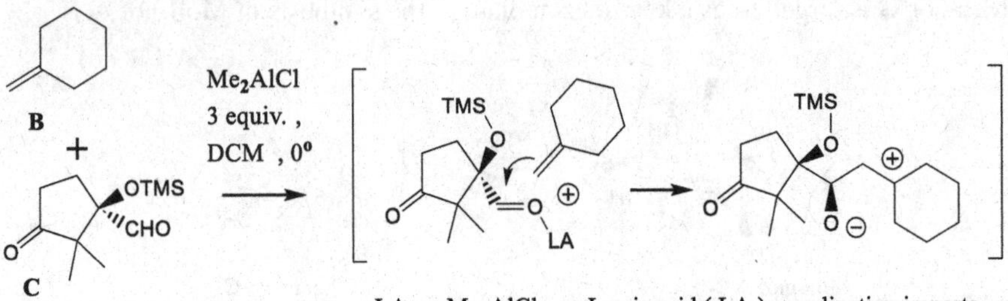

LA = Me$_2$AlCl   Lewis acid (LA) coordination imparts a partial positive charge on the carbonyl group, initiating a nucleophilic attack on it by the alkene. Proton migration proceeds, forming **D**.

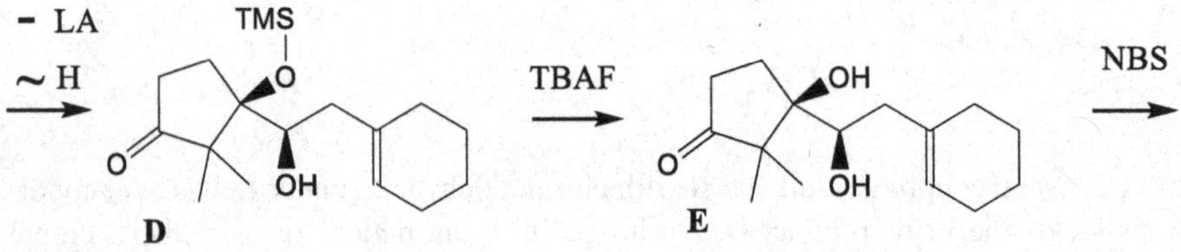

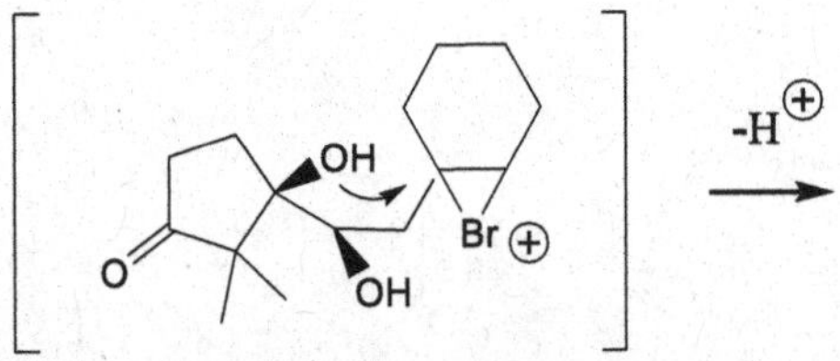

The intermediate bromonium ion is promptly attacked by the tertiary alcohol, closing the central oxa-bispiro ring

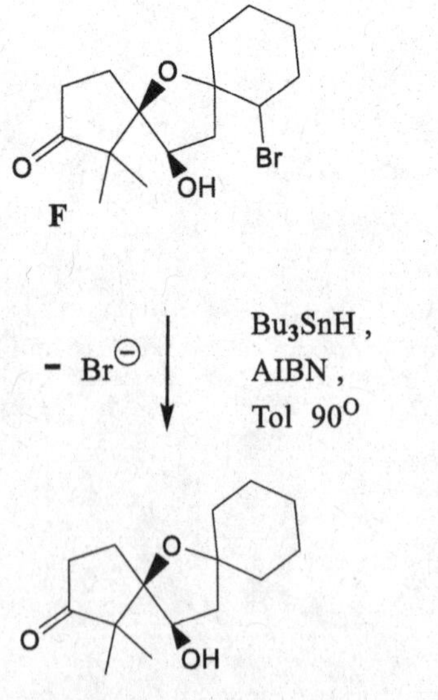

Bu$_3$SnH ,
AIBN ,
Tol 90°

TBAF = tetrabutyl ammonium fluoride
NBS = N-bromo succinimide

Precursor **G**

PREGABALEN

(S) Pregabalen

Gokavarapu, K.; Devkar, R. U.; Gokavarapu, S.; Samala, S. R. K.; Rao, D. P. An efficient and Novel Synthesis of Pregabalin with Enantioenriched Enzymatic Hydrolysis Using CAL-B Enzyme. *Asian Journal of Research in Chemistry* **2019**, 12, 55–57

In this synthesis of (S) pregabalen, two equivalents of cyanoacetamide are used in an aldol-Michael cascade to construct the nitrile-amide intermediate **A**, which was worked up with sulfuric acid resulting in the hydrolysis of both nitriles and amides in A. Simultaneous decarboxylation produced achiral dicarboxylic acid **B**. Its diester **C** is then desymmetrized with the CAL-B enzyme, establishing the needed chiral center in the half ester **D**. Ammonia then converts **D** to the chiral acid-amide **E**. A Hofmann reaction then delivers (S) pregabalen.

Cyanoacetamide
2 equivalents

Aldol addition ,   dehydration ,   Michael addition
of second equivalent

A

H_2SO_4

prompt hydrolysis
and decarboxylation

Dicarboxyllic
acid

B

$MeOH ,$
H_2SO_4

Diester

C

CAL-B
hydrolase

enzymatic
hydrolysis   Chiral half ester (S)

D

NH_3 , H_2O

Chiral acid amide (R)

E

Hofmann
Reaction

DIPEA = diisopropylethylamine

(S) Pregabalen

SPHINGOSINE ANALOGS

OH
HO
CH3
NH2

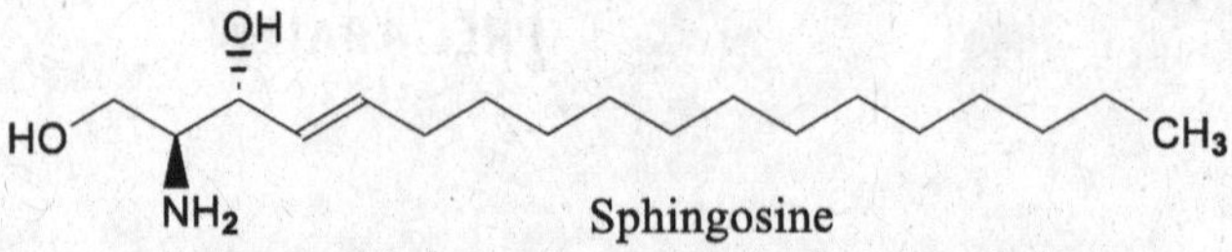

Sphingosine
(2S,3R,4E)-2-Amino-4-octadecene-1,3-diol ;
OR (E)-D-*erythro*-4-octadecene-2(S)-amino-1,3-diol

Saied, E.M.; Le, T.L.-S.; Homemann, T.; Arene, C. Synthesis and characterization of some atypical sphingoid bases. *Bioorg. Med. Chem.* **2018**, *26*, 4047–4057

Four analogs of sphingosine that have been encountered as rare minor metabolites are prepared as shown below.

OH
HO
R
NH2

In sphingosine itself :

R =
$(CH_2)_{13}H$

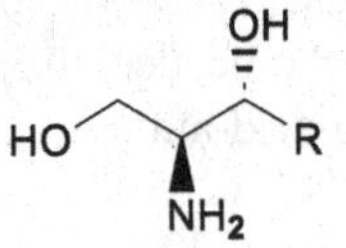

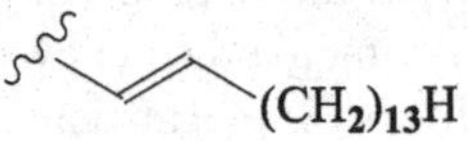

The terminal end of Sphingosine itself or the four analogs.

In the four analogs described here R is a straight chain hydrocarbon, or one functionalized with double bonds and / or a hydroxyl group.

I R = C_{15} Saturated Straight Chain D

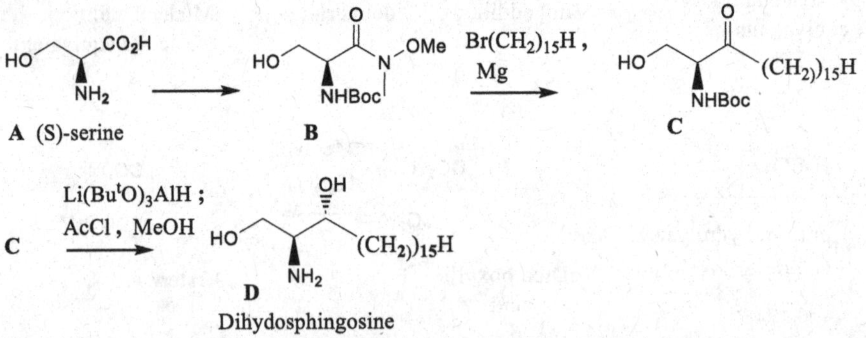

HO
CO2H
NH2
A (S)-serine
O
HO
OMe
N
NHBoc
B
Br(CH2)15H ,
Mg
O
HO
(CH2)15H
NHBoc
C
C
Li(ButO)3AlH ;
AcCl , MeOH
OH
HO
(CH2)15H
NH2
D
Dihydosphingosine

II R = C_{15} with One Misplaced Double Bond F

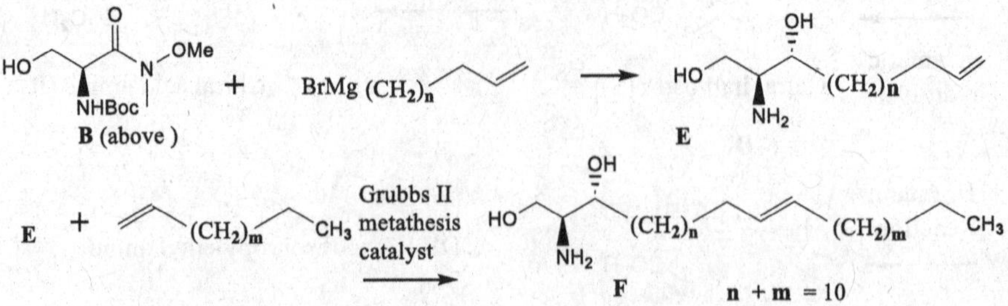

O
HO
OMe
N
NHBoc
B (above)
+
BrMg (CH2)n
OH
HO
(CH2)n
NH2
E
E
+
(CH2)m
CH3
Grubbs II
metathesis
catalyst
OH
HO
(CH2)n
(CH2)m
CH3
NH2
F
n + m = 10

III R = C₁₅ WITH A TERMINAL HYDROXYL K

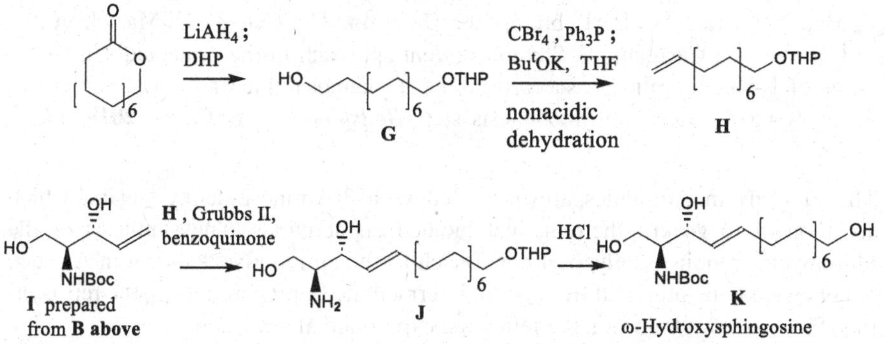

G

H

I prepared
from **B** above

J

K
ω-Hydroxysphingosine

IV R = C₁₅ (VARIABLE) WITH TWO DOUBLE BONDS, ONE E AND ONE Z F

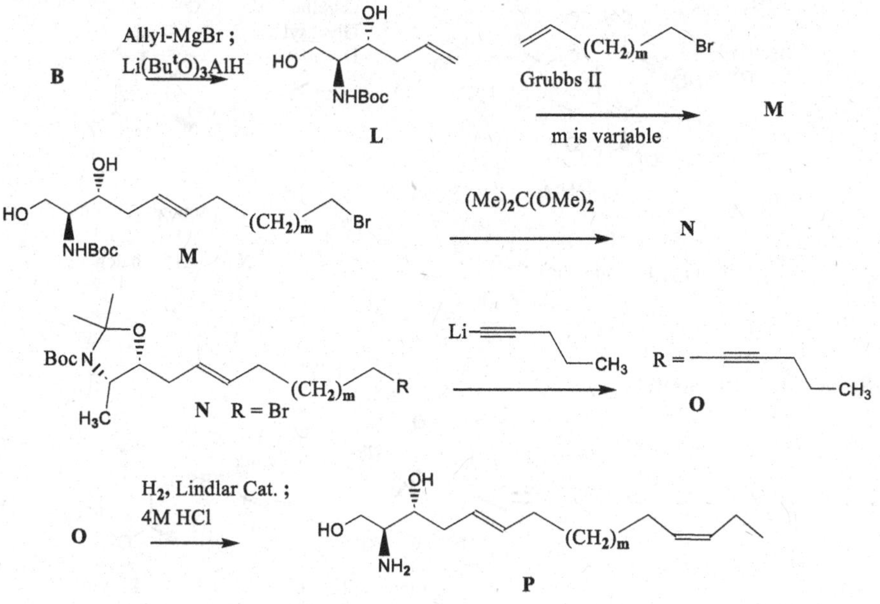

B

L

M
m is variable

M

N

N R = Br

O R =

O

P

DHP = dihydropyran

VANCOMYCIN AMINOGLYCAL

Mace, A.; Legros, F.; Lebreton, J.; Dujardin, G.; Cellet, S.; Martel, A.; Carbone, B.; Carreaux, F. Stereodivergent approach in the protected synthesis of L-vancosamine, L-saccarosamine, L-daunosamine and L-ristosamine involving a ring-closing metathesis step. *Beilstein J. Org. Chem.* **2018**, *14*, 2949–2955

The Pluromycin antibiotics are decorated with 3-Amino-2-deoxy sugars, which are thought to govern their distinct biomedical activities. The synthesis of the Pluromycins requires a glycosyl donor such as the amino glycal shown in A. Such donors were to be prepared from target intermediate B, prepared using the retrosynthetic route shown below, in the following experimental sequence.

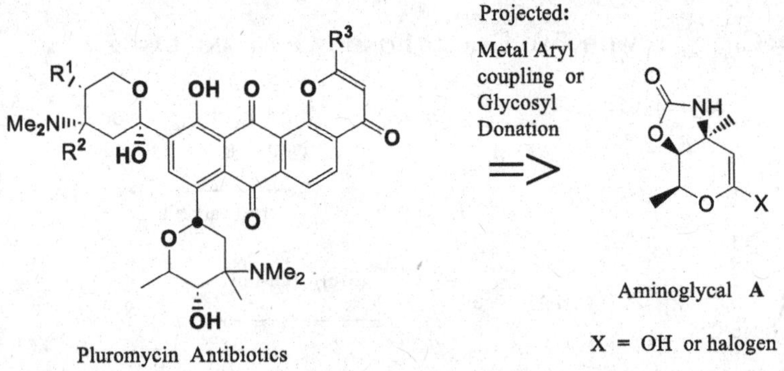

Pluromycin Antibiotics

Projected:

Metal Aryl coupling or Glycosyl Donation

Aminoglycal **A**

X = OH or halogen

RETROSYNTHETIC ROUTE

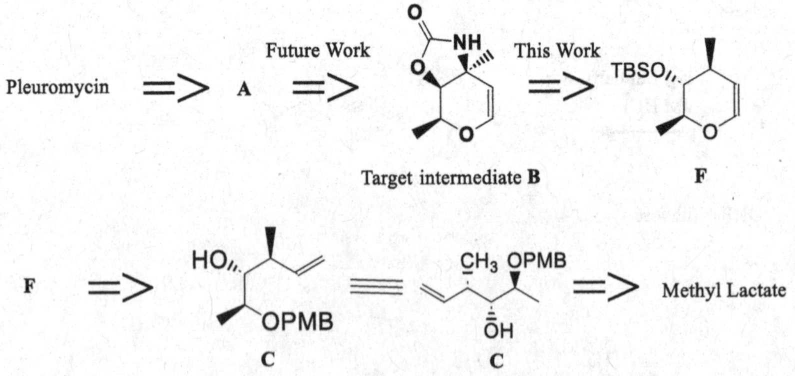

Target intermediate **B**

Methyl Lactate

EXPERIMENTAL SEQUENCE

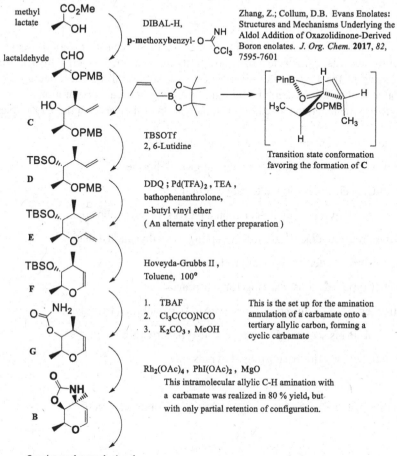

Preparation of the protected lactaldehyde from methyl lactate is reported in :

methyl lactate — CO₂Me / OH

DIBAL-H,
p-methoxybenzyl-

Zhang, Z.; Collum, D.B. Evans Enolates: Structures and Mechanisms Underlying the Aldol Addition of Oxazolidinone-Derived Boron enolates. *J. Org. Chem.* **2017**, *82*, 7595-7601

lactaldehyde — CHO / OPMB

C — HO / OPMB / vinyl

Transition state conformation favoring the formation of **C**

D — TBSO / OPMB

TBSOTf
2, 6-Lutidine

E — TBSO / OPMB / O-vinyl

DDQ ; Pd(TFA)₂ , TEA ,
bathophenanthrolone,
n-butyl vinyl ether
(An alternate vinyl ether preparation)

F — TBSO / O ring

Hoveyda-Grubbs II ,
Toluene, 100°

G — O / NH₂ / O ring

1. TBAF
2. Cl₃C(CO)NCO
3. K₂CO₃ , MeOH

This is the set up for the amination annulation of a carbamate onto a tertiary allylic carbon, forming a cyclic carbamate

B — O / NH / O ring

Rh₂(OAc)₄ , PhI(OAc)₂ , MgO
This intramolecular allylic C-H amination with a carbamate was realized in 80 % yield, but with only partial retention of configuration.

Ongoing work towards **A** and the antibiotics themselves

5 Name Reactions

The standard reactions of the twentieth century have been in use for so long that their mechanisms and purview of applicability are well understood. This has led to a widening of their utility in preparing difficult structural moieties, being used in combinatorial syntheses, and in microgram to kilogram scale preparations.

Here are the reactions listed, with an indication of their uses in these examples:

Tiffeneau–Demjanov: with a 1,2 carbonyl shift

Meyer–Schuster: synthesis of a plant growth hormone

Aza–Claisen: substituted tetrahydrofurans

Petasis: butyrolactam annealed on a bicyclic heterocyclic core

Calvin rearrangement and Stille coupling: methyl carlactonate

Robinson: different types of products

Stetter type reactions: a range of substrates

Strecker: cyclic aminonitriles – preparation, stability, hydrolysis

Claisen–Prins sequence: a propellane type structure

Achmatowitz: for trisubstituted pyrans

Six successive name reactions: Lyngbyatoxin

Stetter-Michael sequence;   new pathways to common targets

DOI: 10.1201/9781003397816-5

UNDESIRED PRODUCT

Lohoelter, C.; Schollmeyer, D.; Waldvogel, R.S.; Derivatives of (-)-isosteviol with expanded ring D and various oxygen functionalities. *Europ. J Org. Chem.* **2012**, *2012*, 6364–6371

HIGHLIGHTS:

- Isosteriol derivatives
- Four derivatives of the natural product are prepared
- Meerwein–Wagner rearrangement
- Tiffeneau–Demjanov (T–D) rearrangement produces unexpected product
- Unexpected product converted into expected product, using a 1,2 carbonyl shift sequence

Isosteviol R = H   **A**
          R = Me **B**

G Expanded Ring D
    product required

FUNCTIONALIZATION TO THE ALDEHYDE

Corey-Chaykovsky
KOBu^t, Me$_3$SI,
DMF

B $\longrightarrow$ **C** Used in T-D rearrangement stage, next page.

Meerwein-Wagner
a: BF$_3$•OEt$_2$, DCM
OR b: LiI , DMSO
OR c: F$_3$CCO$_2$H , DCM

C $\longrightarrow$

Alternate conversion
methods yield only
the aldehyde **D**

D This aldehyde was carried no furthur

TIFFENEAU–DEMJANOV REARRANGEMENT

C → NaN₃ , DMF 90° → E → H₂ , Pd/C → F

F → NaNO₂, HOAc, 0°
Tiffeneau-Demjanov
rearrangement → G Expected 0 %   H Unexpected 67 %

CONVERSION OF UNEXPECTED PRODUCT H TO DESIRED PRODUCT G WITH A 1,2 CARBONYL SHIFT SEQUENCE

H → NBS , TsOH 145° → I

I → NaOH , DMF → J + K

K → NaBH₄ , MeOH → L 81%

L → Deoxygenation TMSI , DCM 58% → G Expected Tiffeneau-Demjanov product

ADDENDUM

Wu, Z.; Xu X.; Wang, J.; Dong, G.; Carbonyl 1,2-transposition through triflate-mediated α-amination. *Science* **2021**, #6568, 734–740

An alternate method of realizing such a 1,2 carbonyl shift has been reported. Dihydrotestosterone acetate was converted to its carbonyl shifted analog as illustrated below. The triflate was amminated, deoxygenated to the enamine, and spontaneously hydrated with a palladium norbornene catalyst and a wet amine.

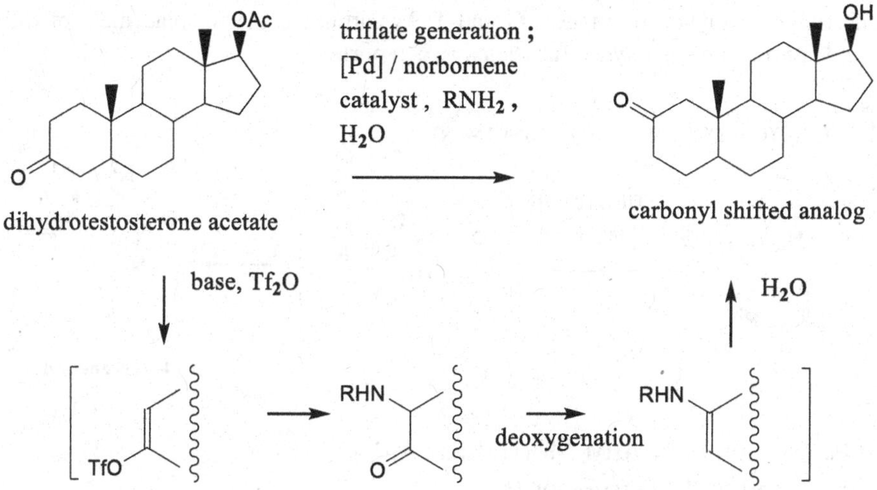

dihydrotestosterone acetate

triflate generation ;
[Pd] / norbornene
catalyst , RNH$_2$,
H$_2$O

carbonyl shifted analog

base, Tf$_2$O

H$_2$O

deoxygenation

Compare with the conversion of unexpected T-D product **H** to **G** on previous page.

AZA-CLAISEN REARRANGEMENT

"Five examples."

Davidson, S.J.; Pilkington, L.I.; Dempsey-Hibbert, N.C.; El-Mohtadi, M.;
Tang, S.; Wainwright, T.; Whitehead, K.A.; Barker, D. Modular Synthesis and
Biological Investigations of 5-Hydroxymethyl Dibenzyl Butyrolactones and
Related Lignins. *Molecules* **2018**, *23*, 3057

Claisen rearrangement preparation of 4-pentenamides (**D**) and their conversion into
substituted tetrahydrofuranones (**G** and **J**) by intramolecular displacement of the
amide nitrogen by an oxygen functionality is reported.

I THE ACYL AZA-CLAISEN REARRANGEMENT

II PREPARATION OF N-ALLYL MORPHOLINES A, AND CONVERSION TO 4-PENTENAMIDE D

W , V are various substituents

OsO$_4$, NMO , Bu^{t}OH , H$_2$O

NaIO$_4$

$\overset{\oplus}{P}h_3P\overset{\ominus}{-CH}-CO_2H$

DIBAL-H , DCM , -78°

MsCl , TEA , Morpholine

NMO = N-Methylmorpholine-N-oxide

Acyl-Aza
Claisen

TiCl$_4$•2THF,
Hunig Base   **D** 4-Pentenamide

III Tetrahydrofuranone Preparation from Pentenamides of Type D

D $R^1 = Bn^1$
        $R^2 = Bn^2$

Product chirality is determined by method of workup.

IV Tetrahydrofuran

D $R^1 = R^2 = $ Phenyl

V Tetrahydrofuran

G
From Example 1 in **III**

L   **M**

PETASIS REACTION: ANGULAR
HETEROSPIROCYCLE

Flagstad, C.M.G.; Azevedo, G.; Min, A.; Willaume, R.; Morgentin, R.; Nielsen, T.E.; Clausen, M.H. Petasis/Diels–Alder cyclization cascade reactions for the generation of scaffolds with multiple stereogenic centers and Orthogonal handles for library production. *Europ. J ofOrg. Chem.* **2018,** *2018.* 5023–5029

The synthesis of polycyclic spirocycle scaffolds with three stereocenters, a high fraction of sp^3 carbon atoms, low molecular weight, and three handles for functional diversification is described. The one reaction to generate the key intermediate **G** is an imine formation, followed by a Diels–Alder reaction, followed by an intramolecular Petasis cyclization as follows.

As the three starting materials interact, the imine **D** promptly reacts with the furan boronic acid **C** to form the Diels–Alder bicyclic intermediate **E**. Intramolecular boronic acid Mannich coupling (the Petasis reaction) of **E** then generates the tricyclic amino etheric acid **F**. The second nitrogen is introduced with amide **G** formation.

TBTU = O-(Benzotriazole-l-yl)-N,N,N',N'-tetramethyluronium tetrafluoroborate

Oxidative/reductive elaboration of intermediate **G** proceeds as shown below.

The resultant diol **I** is cyclized to the spirocycle. Combinatorial synthesis could be made for the introduction of substituents R^1, R^2, and R^3.

Compounds of such complexity have shown promise as therapeutic agents, and the concise synthesis shown is scalable.

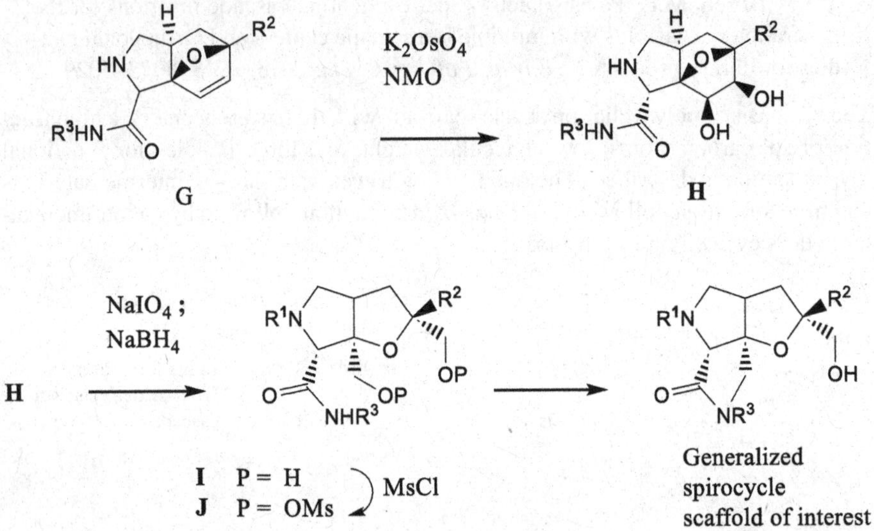

$$K_2OsO_4$$
$$NMO$$

G   **H**

H $\xrightarrow{\begin{array}{c}NaIO_4\,; \\ NaBH_4\end{array}}$

I P = H
J P = OMs $\Big\}$ MsCl

Generalized
spirocycle
scaffold of interest

NMO = N-Methyl morpholine-N-oxide

METHYL CARLACTONATE

Dieckmann, M.C.; Dakas, P.-Y.; DeMesmaeker, A. Synthetic access to noncanonical strigolactones: Synthesis of carlactonic acid and methyl carlactonate. *J. Org. Chem.* **2018** *83*, 125–135

HIGHLIGHTS:

- Oxy–Michael reaction of a hemiacetal on ethyl propiolate (**B** to **C**)
- α-Iodination of the resulting β-oxa-acrylate (**C** to **D**)
- Calvin's method to convert aldehyde **E** to acetylene **F**
- Modified free radical procedure for the hydrostannylation of acetylene **F**
- A Still coupling realized the strategic disconnect below under unconventional dual-arsenic cofactor conditions for the coupling of Iodide **D** with tributyl stannane **G**

Methyl Carlactonate **A**

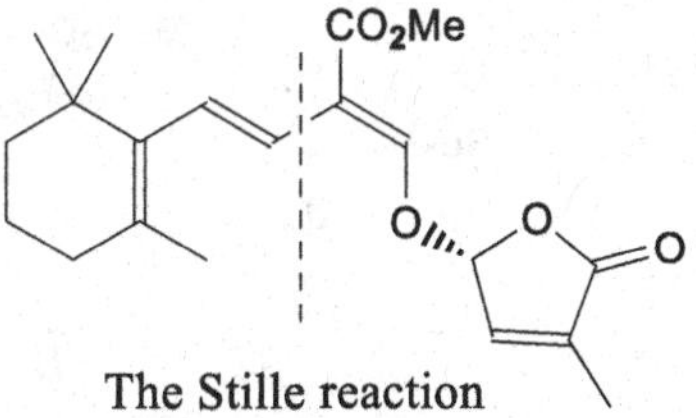

The Stille reaction disconnect

The synthesis of the triene is dictated by the Stille disconnect above, using the coupling of an iodide **D** with a tributyl stannane **G**.

I THE STILLE IODIDE PREPARATION FROM AN ALKYNE

This efficient preparation of the α-iodo-β-oxa-acrylate (**D**), a key intermediate in the carlactonate synthesis, is prompted by the disconnect shown above.

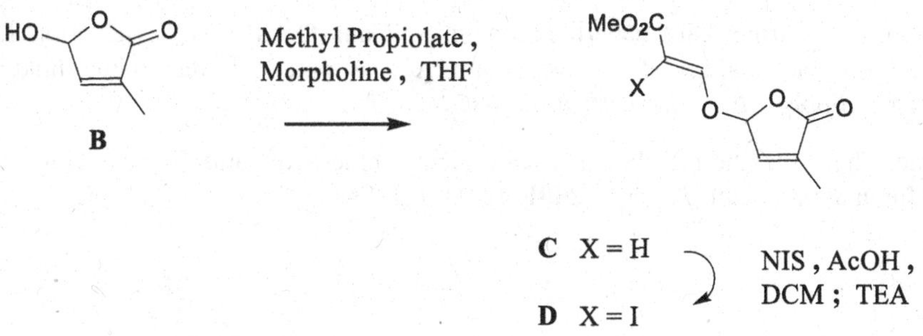

B

Methyl Propiolate ,
Morpholine , THF

$\longrightarrow$

C X = H
D X = I

NIS , AcOH ,
DCM ; TEA

II THE STILLE TRIBUTYL STANNANE PREPARATION FROM A CALVIN ENE-YNE INTERMEDIATE

In this reaction, the Calvin approach is an efficient alternative to the Ohira–Bestman reagent.

III THE STILLE COUPLING

For other examples of the Calvin rearrangement, see:

Odlo, K.; Fournier-Dit-Chabert, J.; Ducki, S.; Gani, O.A.B.S.M.; ... Hansen, T.V. 1,2,3-Triazole analogs of combretastatin A-4 as potential microtubule-binding agents. *Bioorg. Med. Chem.* **2010**, *18*, 6874–6885

Kao, Chai-Lin; Chern, Ji-Wang A convenient synthesis of naturally occurring benzofuran ailanthoidol. *Tet. Lett.* **2001**, *42*, 1111–1113

ROBINSON REACTION PRODUCTS I, II, AND III

I ROBINSON REACTION WITH MULTIPLE POSSIBLE PRODUCTS

Ramachary, D.B.; Reddy, P.S.; Gujral, J. Construction of 2-Thiabicyclo[3.3.1] nonanes by organocatalytic asymmetric formal [3+3] cycloaddition. *Europ. J Org. Chem.* **2018**, *2018*, 1852–1857

A reaction in which Robinson conditions could deliver three cyclic products.

The kinetic enolate formed by the initial Michael reaction of **A** with **B** is shown below as the Michael enolate rearranging into the Robinson and tertiary enolates. Successive hydrogen shifts rearrange this to the equilibrium forms shown as the Robinson enolate and the tertiary enolate.

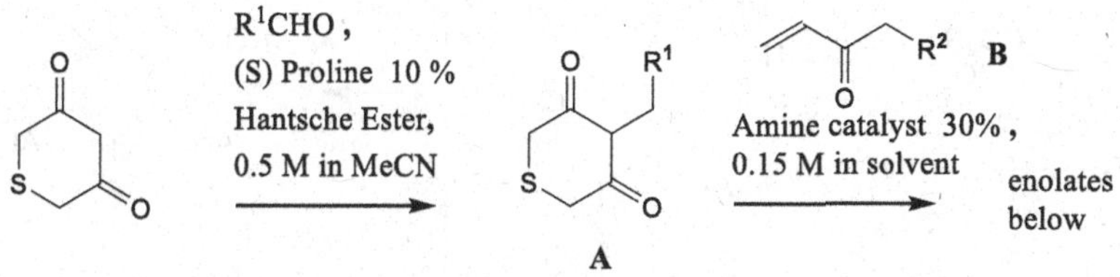

A

Each of the three enolates in this equilibriun could cyclize in a different way.

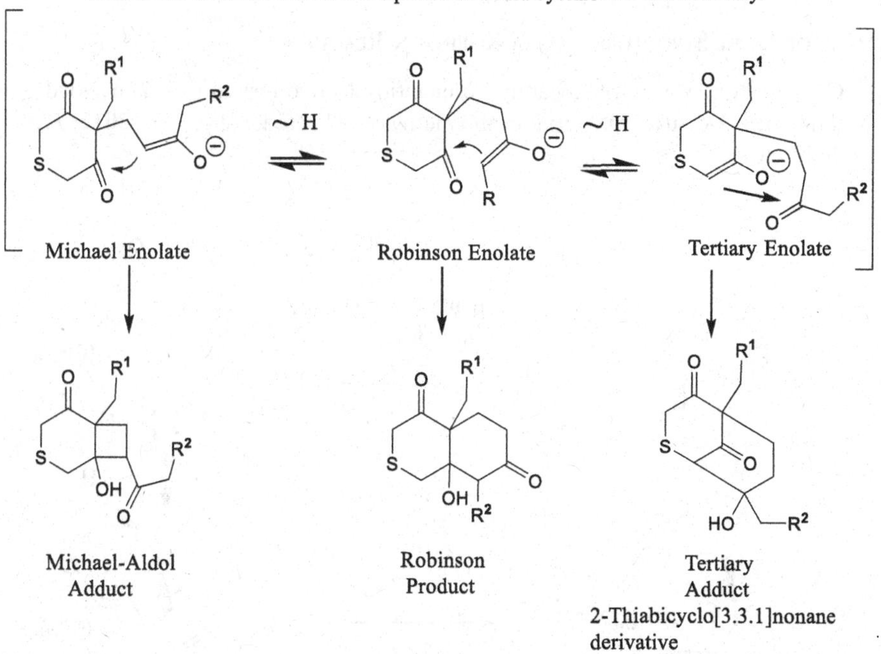

Michael Enolate   Robinson Enolate   Tertiary Enolate

Michael-Aldol
Adduct

Robinson
Product

Tertiary
Adduct
2-Thiabicyclo[3.3.1]nonane
derivative

II An example of the Reaction Giving a Single Monocyclic Product Is Given Below

Zhao,W.; Zhamg, X.; Liu, M.; Vasconcelos, S.N.S.; Zhang, W. One-pot Fluorination and Organocatalytic Robinson Annulation of Mono- and Difluorinated Cyclohexenone. *Molecules* **2018**, *23*, 2251

III Propellane Structure from a Robinson Reaction

Cai, Q.; You, S.-L. Organocatalyzed enantioselective formal [4 + 2] cycloaddition of 2,3-disubstituted indole and methyl vinyl ketone. *Org. Lett.* **2012**, *14*, 3040–3043

Propellane derivatives of tetrahydrocarbazole

NHC CATALYSED ALDEHYDE COUPLING

Gaggero, N.; Pandini, S. Advances in chemoselective intermolecular cross-benzoin-type condensation reactions. *Org. Biomol. Chem.* **2017**, *15*, 6867–6887

N-Heterocyclic Carbenes (NHC) such as **A** catalyse the coupling of aldehydes to electrophiles. **Three examples of this include:**

I BENZOIN CONDENSATION DIMERIZATION OF ALDEHYDES

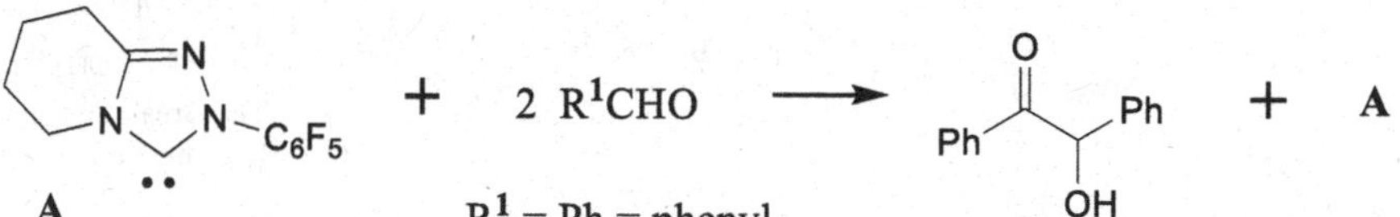

$R^1 = Ph = phenyl$

benzoin

II STETTER REACTION: CROSS-BENZOIN COUPLING
COUPLING OF DISSIMILAR ALDEHYDES

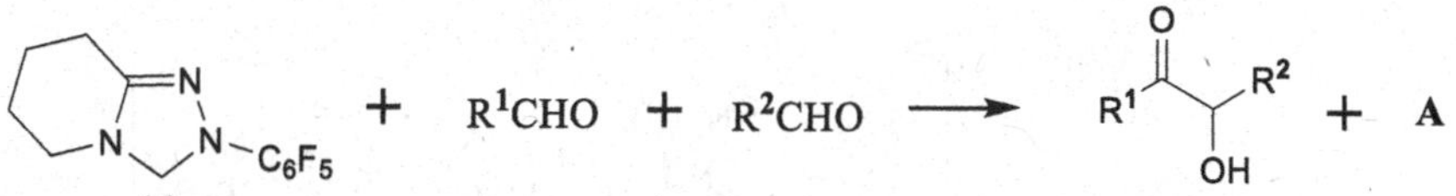

R^1 or R^2 = aromatic or aliphatic

acyloin

III STETTER–MICHAEL REACTION: ADDITION OF AN
ALDEHYDE TO AN A,B UNSATURATED EWG

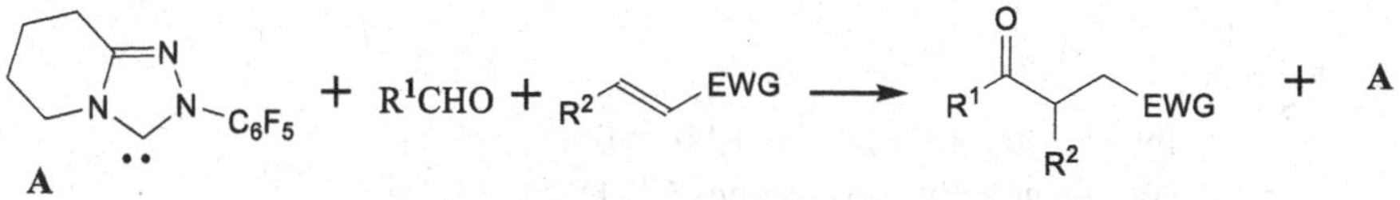

EWG = electron withdrawing group.

The mechanism of these couplings is illustrated as follows.

Each of the three reactions is initiated by the carbene insertion of NHC **A** into the C–H bond of aldehyde R^1CHO, forming a ketone like **B**, which enolizes to **C**.

I BENZOIN CONDENSATION

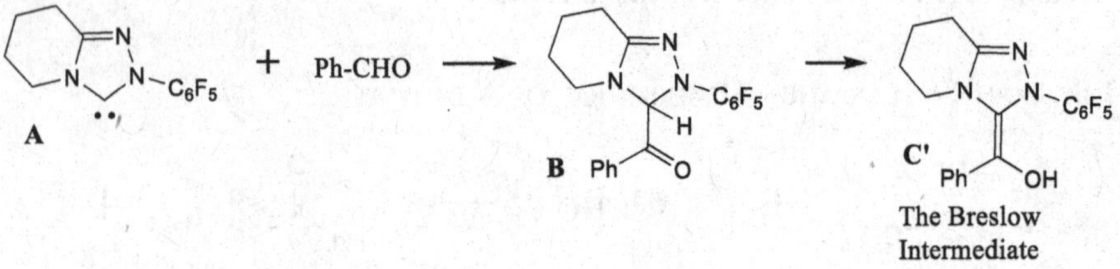

A   +   Ph-CHO   →   **B** Ph   →   **C'**

The Breslow
Intermediate

Breslow proposed that an intermediate such as enoldiamine **C'** is the nucleophile that attacks a second equivalent of benzaldehyde, as shown below. Note that the nitrogens on the enol induce a *umpolung* nucleophilicity to the benzylic carbon of **C**, attacking the aldehyde and forming **D'**. An intramolecular hydrogen shift within alkoxide **D'** produces isomeric alkoxide **E'**, which then expells benzoin and NHC**A**

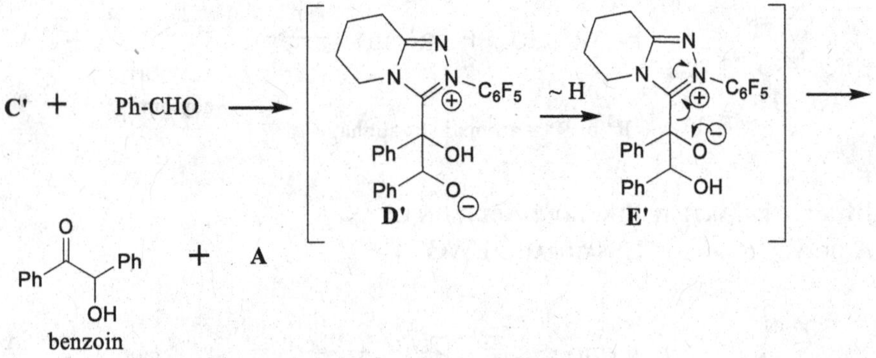

C' +   Ph-CHO   →   **D'**   ~ H   **E'**   →

benzoin   +   **A**

Intermediates **C'**, **D'**, and **E'** in this example are congruent with intermediates **C''**, **D''**, and **E''** in example **II**, and with **C'''**, **D'''**, and **E'''** in example **III**.

II Cross-Benzoin Coupling: The Cross Product of Two Different Aldeydes R¹CHO and R²CHO Begins as Above in I, Then Proceeds as Below

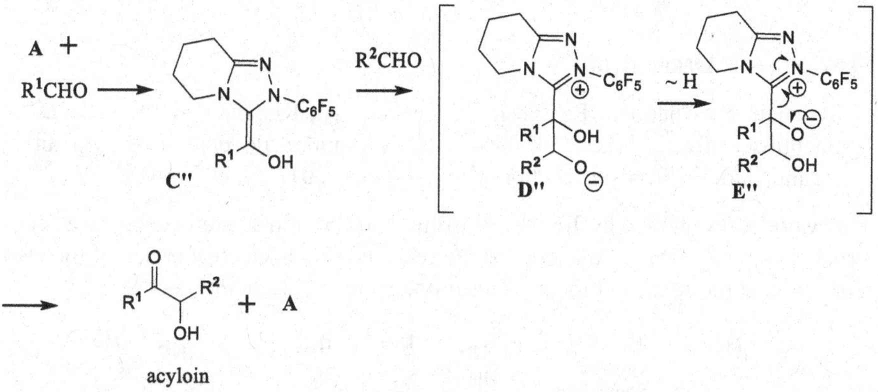

acyloin

III Stetter–Michael Reaction

As in **I** and **II**, the nucleophilic Breslow intermediate **C** is initially formed. Nucleophilic attack of the hydroxyl-bearing carbon of **C** on the α,β unsaturated electron withdrawing group occurs in Michael fashion forming intermediate **J**, which undergoes a hydrogen shift producing **K**, which fragments as shown into the Michael product **L** and NHC **A**.

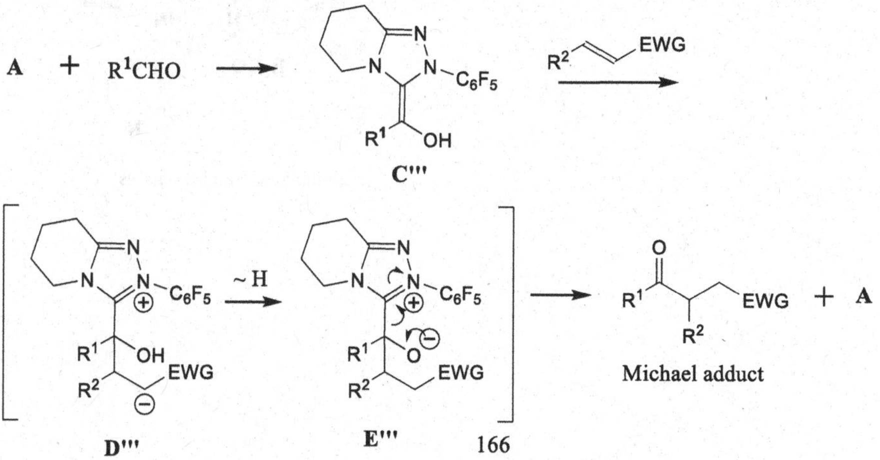

Michael adduct

STRECKER REACTION – CYCLIC AMINONITRILES

Two examples of substituted Strecker products, their formation, stability, and hydrolysis to cyclic aminonitriles.

I PYRAZOLONE DERIVATIVES

Mahajan, S.; Chauhan, P.; Kaya, U.; Deckers, K.; Rissanen, K.; Enders, D. Enantioselective synthesis of pyrazolone α-aminonitrile derivatives via an organocatalytic Strecker reaction. *Chem. Comm.* **2017**, *53*, 6633–6636

This work shows how a cyclic tetrasubstituted amino nitrile stereocenter was constructed using a chiral organocatalyst. Stereo- and regio-selective attack of the cyanide anion at the *re* face of the exterior imine yields the R-enantiomer **B**.

Quinine Squareamide catalyst

II CYCLIC SUBSTITUTED AMINONITRILES

Voznesenskaia, N.G.; Shmatova, O.I.; Sosnova, A.A.; Nenajdenko, V.G. From cyclic ketimines to α-substituted cyclic amino acids and their derivatives: Influence of ring size and substituents on stability and reactivity of cyclic aminonitriles. *Europ. J. Org. Chem.* **2019**, *2019*, 625–632

Synthesis   R = Phenyl, alkyl

D   n = 1,2,3
        R = various

E

F

Cylic Alpha Aminonitrile Stability, variation with ring size

G Forms rapidly, but too unstable to isolate

H Stable

I Slightly unstable

Hydrolysis of aminonitriles H and I

Formyl protection of **I** is required to retain stability prior to hydrolysis. Note the acetic anhydride/formic acid conditions used in this protection of **I**. HCl deprotects the amine and hydrolyses the nitrile.

I   P = H

     P = CHO

OXY-BRIDGED TETRACYCLIC FRAMEWORK

J

Nair, P.N.; Ghaytadak, V.S.; Sridhar, N.B.; Basi, J.G.R.; Reddy, V.S. Constuction of oxa-bridged tetracyclic frameworks through a prins bicyclic annulation. *Europ. J. Org. Chem.* **2019**, *2019*, 3567–3574

This sequence of the two-stage acetylene-allene Claisen-based rearrangement below, and the two-stage Prins reaction (next page), constitute a useful concise approach to dioxypropellane moieties of this type (**J**).

The Claisen Rearrangements

The first stage of the Claisen rearrangement (**B** to **C**) is followed by hydrogen migration (**C** to **D**). In the second stage, intermediate **D** forms **E** which is then reduced to diene diol **F**.

Reduction of the butadiene diacid **E** to diol **F** is followed by the Diels–Alder/Prins sequence shown on the following page.

THE PRINS REARRANGEMENTS

The butadiene diol product **F**, from the Claisen/reduction sequence, is converted into the deoxypropellane **J**, in a Diels-Alder/two-stage-Prins sequence.

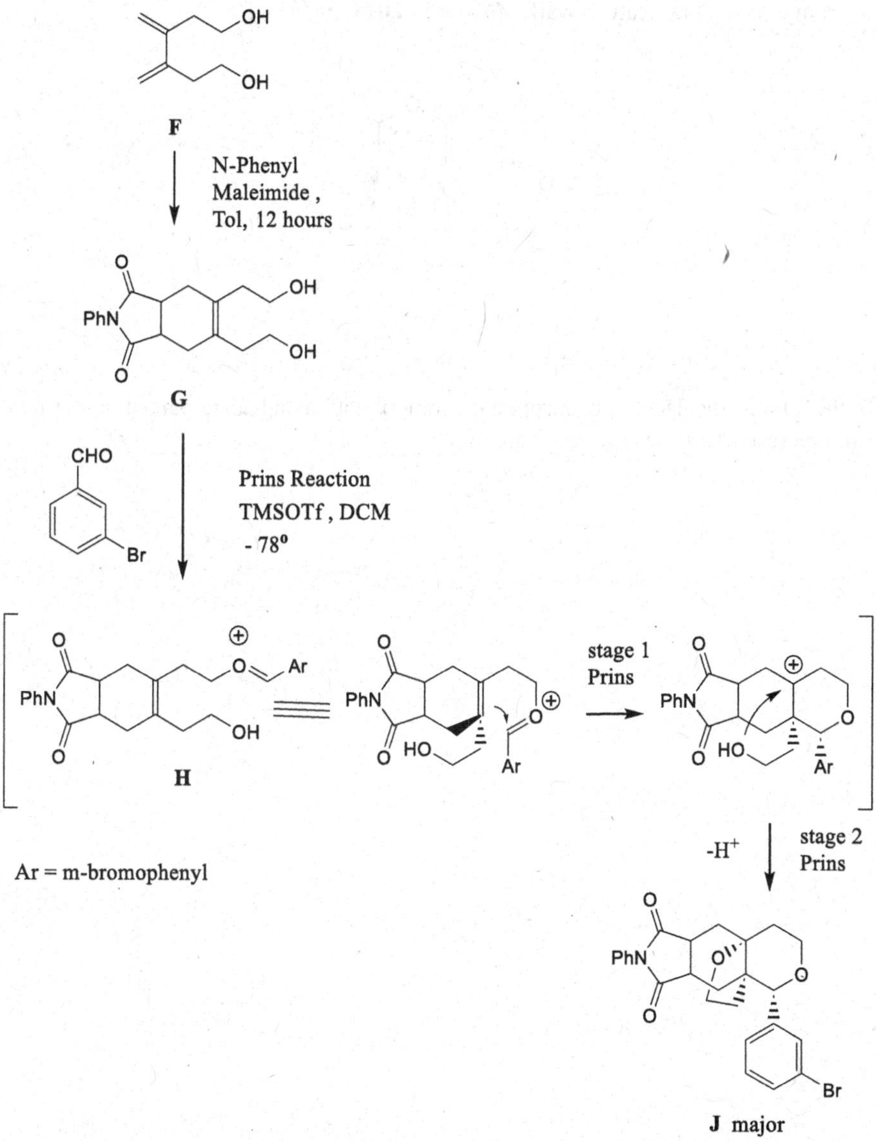

F

N-Phenyl
Maleimide,
Tol, 12 hours

G

Prins Reaction
TMSOTf, DCM
- 78°

H

Ar = m-bromophenyl

stage 1
Prins

-H⁺   stage 2
Prins

J major

J is a derivative of: hexahydro-8a,4a-(epoxyethano)pyrano[3,4]isoindol-1,3(2H,5H)-dione.

ADDENDUM

A structurally related tricyclic moiety occurs in a natural product described in:

Farooq, U.; Pan, Y.; Disasa, D.; Qi, J. Novel anti-aging benzoquino derivatives from onosma bracteaturn wall. *Molecules* **2018**, *24*, #1428

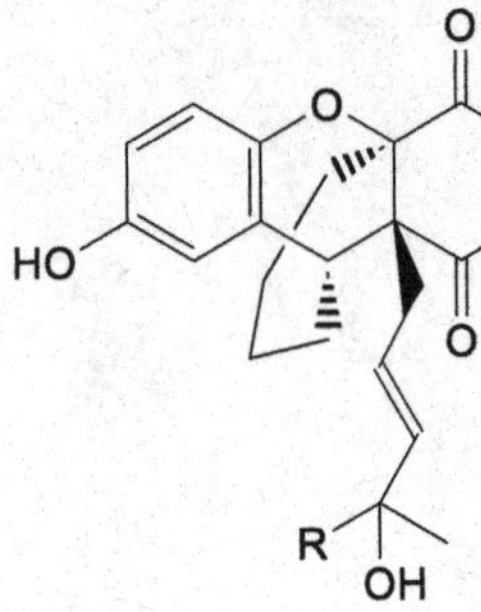

In this case, a displaced dioxapropellane moiety with an adjacent pair of quaternary carbons resembles the core structure of **J**.

ACHMATOWICZ – DECYTOSPOLIDE B

Ghosh, A. K.; Simpson, H. M.; Veltschegger, A. M. Enantioselective total synthesis of decytospolide A and decytospolide B using an Achmatowicz reaction. *Org. Biomol. Chem.* **2018**, *16*, 5979–5986

Decytospolide B exhibits in vitro ctotoxic activity against tumor cell lines A549 and QGY.

Highlights:

- Preparation of a substituted furan
- Assymmetric hydrogen transfer reduction of the ketone (**C** to **D**) for the Achmatowicz starting material
- Rearrangement mechanism of the epoxide intermediate (**D** to **E**) in the Achmatowicz reaction
- Synchonous deoxygenation, alkene saturation, and carbonyl reduction step (**E** to **F**),
- Transformation of **F** into the natural product decytospolide

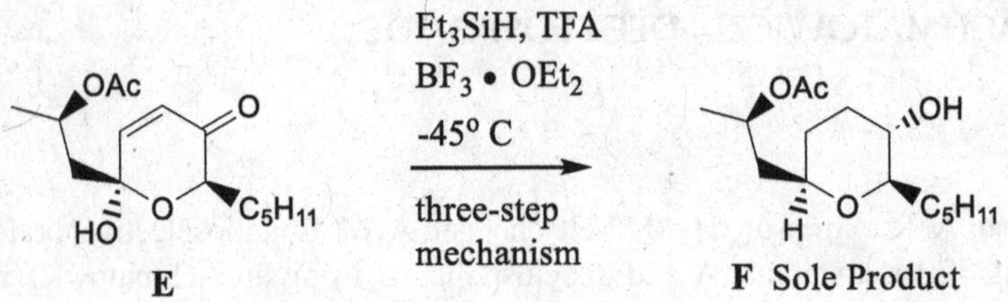

E

Et$_3$SiH, TFA
BF$_3$ • OEt$_2$
-45° C

three-step
mechanism

F Sole Product

This reaction stereoselectively reduces the α,β unsaturated carbonyl to the saturated alcohol, and deoxygenates the tertiary alcohol without epimerization.

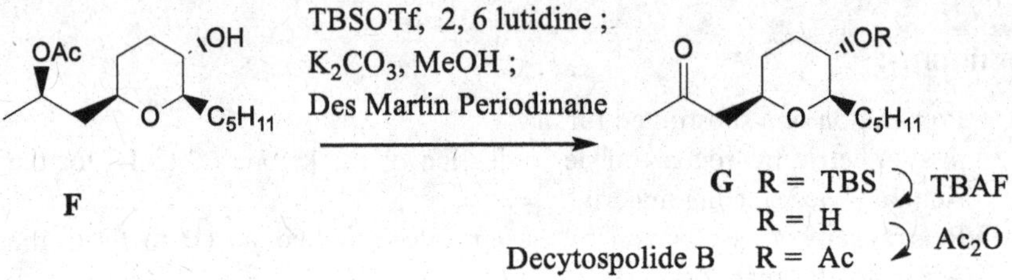

F

TBSOTf, 2, 6 lutidine ;
K$_2$CO$_3$, MeOH ;
Des Martin Periodinane

G   R = TBS ⟩ TBAF
     R = H
Decytospolide B   R = Ac ⟩ Ac$_2$O

THE QUATERNARY CENTER OF LYNGBYATOXIN A

Vitgal, P.; Tanner, D. Efficient and highly enantioselective formation of the all-carbon quaternary stereocenter of lyngbyatoxin A. *Org. Biomol. Chem.* **2006**, *4*, 4292–4298

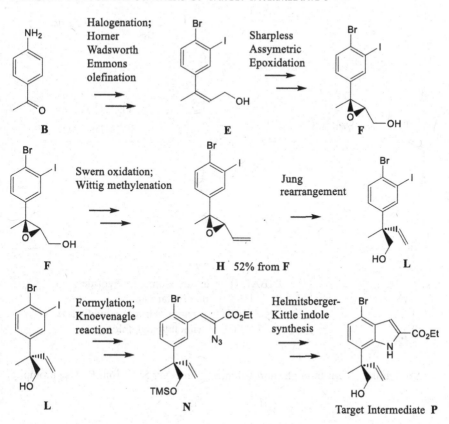

Lyngbyatoxin-A

Projected future synthesis

Intermediate **P**, target of this synthesis

OVERALL PLAN FOR THE SYNTHESIS OF TARGET INTERMEDIATE P

B — Halogenation; Horner Wadsworth Emmons olefination → **E** — Sharpless Assymetric Epoxidation → **F**

F — Swern oxidation; Wittig methylenation → **H** 52% from **F** — Jung rearrangement → **L**

L — Formylation; Knoevenagle reaction → **N** — Helmitsberger-Kittle indole synthesis → Target Intermediate **P**

PREPARATION OF THE JUNG REARRANGEMENT STARTING MATERIAL H

The quaternary center of interest in **P** is to be formed from intermediate **H**. A Sharpless Assymetric Epoxidation (**E to F**, in quantitative yield and high diastereomeric excess) generates the chiral epoxide **F**, in the conformation needed in **H** for the forthcoming Jung rearrangement, described on the next page.

B → (1) ICl , CaCO$_3$, MeOH (2) NaNO$_2$, HBr , -10° ; CuBr , HBr 80° → **C**

C → EtO$_2$C—CH$_2$—P(OET)$_2$ (O) , NaH , THF , 0° →

D → DIBAL-H , Et$_2$O → **E**

E → L(+) DET , Ti(OPr)$_4$, Bu^{t}OOH , DCM , 100% , 92% de →

F → Swern [O] → **G**

G → Ph$_3$P-CH$_3$, KHMDS , THF →

H

DIBAL-H = dibutylaluminum hydride
DET = diethyl tartrate
KHMDS = potassium hexamethyldisilazide
TMSCl = trimethylsilyl chloride

H Labile. Filter through short column and recover 52% from **F**. Use promptly.

JUNG REARRANGEMENT AND FORMATION OF THE INDOLE MOIETY

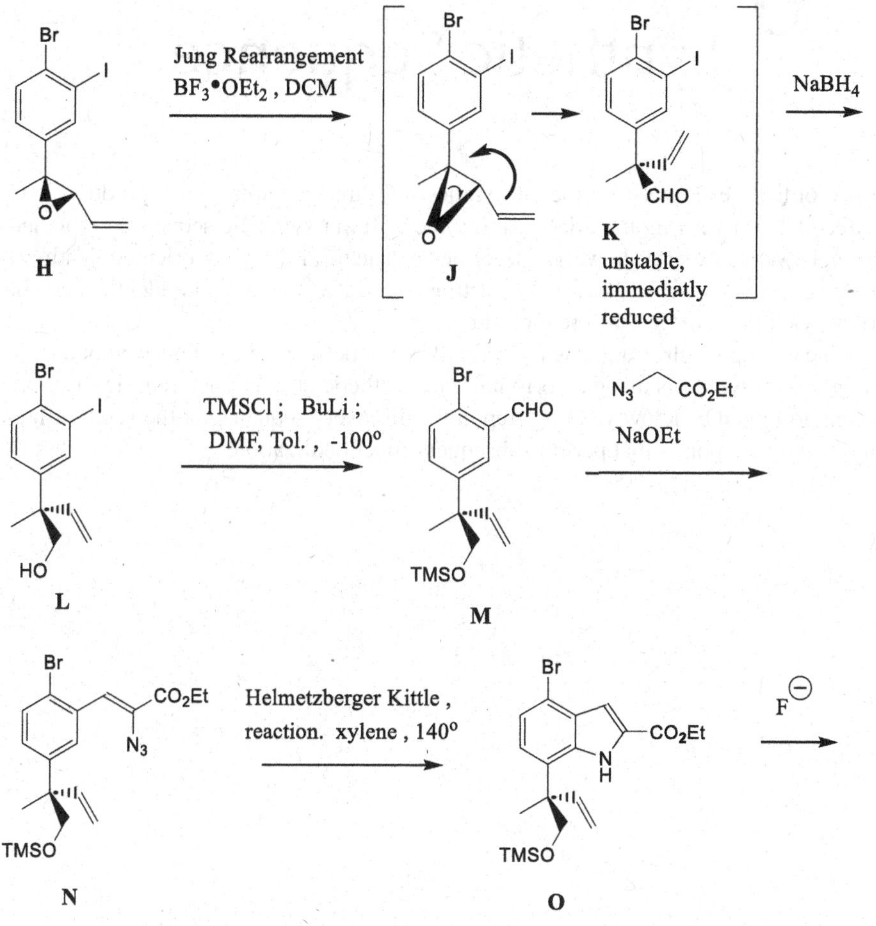

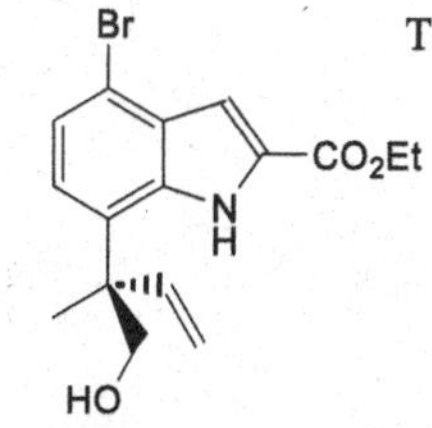

H → **Jung Rearrangement** BF₃•OEt₂ , DCM → [J → K unstable, immediatly reduced] → NaBH₄

L → TMSCl ; BuLi ; DMF, Tol. , -100° → M

M → N₃—CO₂Et / NaOEt →

N → **Helmetzberger Kittle , reaction. xylene , 140°** → O → F⁻ →

The total synthesis of lyngbyatoxin was ultimately documented in 2014:

Nathel, N.F.F.; Shah, T.K.; Bronnera, S.M.; Garg, N.K.
Total synthesis of indolactam alkaloids (-)-indolactam V,
(-)-pendolmycin, (-)-lyngbyatoxin A, and (-)-teleocidin A-2.
Chemical Science **2014**, 2184-2190

Target Intermediate **P**

6 Variations on a Synthetic Sequence

Most of these excerpts describe the synthesis of three or fewer related products with three or fewer starting materials using three or fewer synthetic sequences. The heteropolycyclics excerpt, however, describes a structural diversity-oriented synthesis on an exceptional scale: 14 pairs of starting materials, reacted under identical conditions, yielding fourteen different products.

The examples illustrate how an optimal synthetic scheme can be developed from a discovery synthesis using an original retrosynthetic design. Such redesign is more often prompted by a low-yielding step in the discovery sequence or the need for facile attachment points that permit subsequent functionalization.

DOI: 10.1201/9781003397816-6

TRICLODIC ACIDS

Heard, D.M.; Tyler, E.R.; Cox, R.J.; Simpson, T.J. Structural and synthetic studies on maleic anhydride and related natural products. *Tetrahedron* **2020**, *76*, #130717

Wood-rotting pathogens account for significant losses in the lumber industries. Certain fungal natural products such as sytalin and isoglaucanic acid inhibit the growth of these pathogens. These two natural products are thought to arise biosynthetically from triclodic acid precursors. In a long term study, three of these acids were prepared for investigation of the process. Allylic bromide **I** is the key synthetic precursor to the diene stem of these acids.

Synthetic Outline

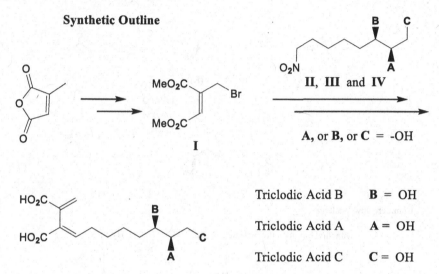

Triclodic Acid B   **B = OH**

Triclodic Acid A   **A = OH**

Triclodic Acid C   **C = OH**

DIMETHYL BROMOMETHYL MALEATE **I** PREPARATION

PREPARATION OF NITROALCOHOLS, AND CONDENSATION
WITH DIMETHYL BROMOMETHYL MALEATE I

Triclodic Acid **C**   Triclodic Acid **A**   Triclodic Acid **B**

HETEROPOLYCYCLICS

Jia, X.; Li, P.; Liu, X.; Lin, J.; Chu, Yu. J.; Wang, J.; Liu, H.; Zhao, F. Green and facile assembly of diverse fused n-heterocycles using gold-catalyzed cascade reactions in water *Molecules* **2019**, *24*, #988

This reaction was used to prepare a library of analogs. In planning for the syntheses, the starting materials, mechanism, and expected product were represented as follows. The analogs would use different chain lengths (**n** and **m**), benzylated or partially benzylated analogs, and hetero-substituted or cyclic moieties in place of the X–Y group.

$n = 0, 1$
$R = H, CH_3$

$m = 0, 1$
$X = C, N, O$
$Y = N, C$

Within the amines β, X is an active atom with an exchangeable hydrogen attached. The X–Y pair may be a heteroatomic bond or part of a cyclic structure.

The mechanism begins with the gold catalyzed carboxylation of the alkyne, as in the Meyer–Schuster reaction, This is followed by tautomerization of the enol to the ketone, imine formation, and coupling, which closes the ring.

The amines β, composed from structural iterations on the X–Y moiety, were cho-sen from the group below, including benzannulated versions and homologs with **n** and **m** equal to 0 or 1. A partial listing of the starting materials employed to synthe-size the products in the library is shown on the next pages.

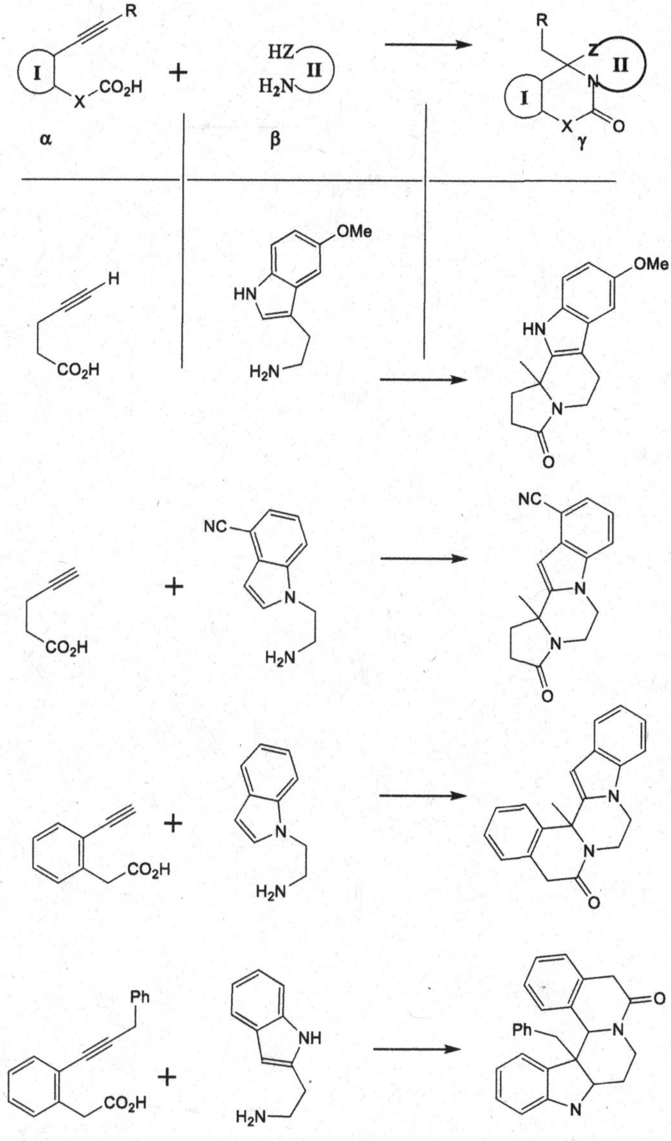
R
I
CO_2H
X
+
HZ
II
H_2N
α
β
R
Z
II
N
I
X
O
γ
H
CO_2H
OMe
HN
H_2N
OMe
HN
N
O
CO_2H
+
NC
N
H_2N
NC
N
N
O
+
CO_2H
N
H_2N
N
N
O
Ph
+
CO_2H
NH
H_2N
O
Ph
N
N

A

B

CH4

D

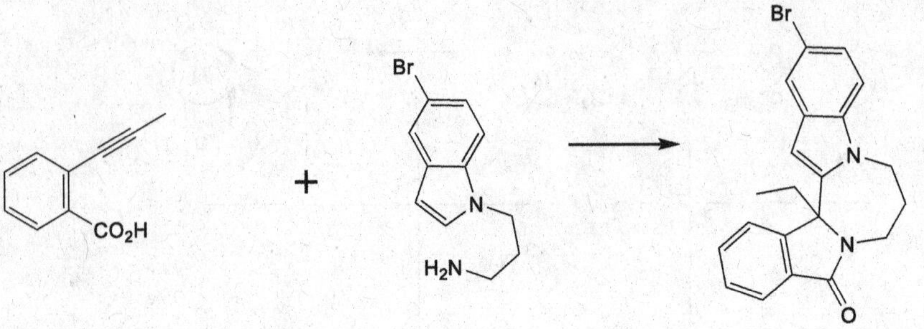
CO2H
+
Br
N
H2N
Br
N
N
O

E

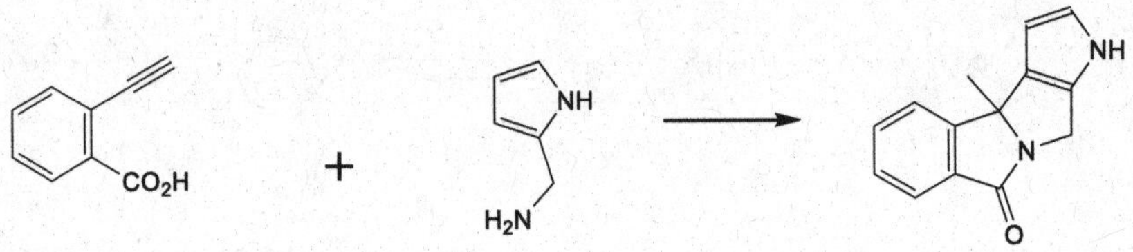
CO2H
+
NH
H2N
NH
N
O

F

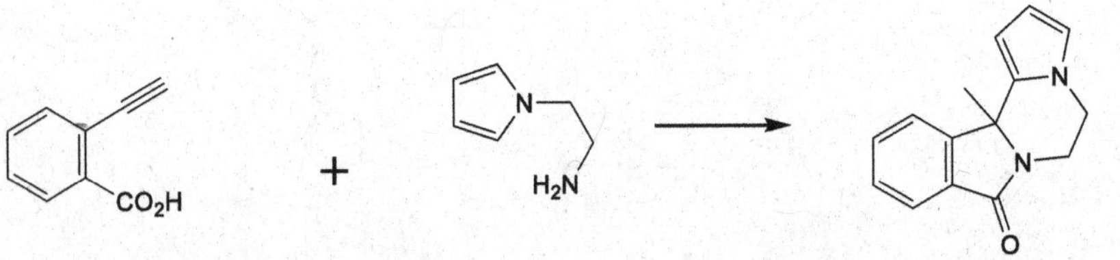
CO2H
+
N
H2N
N
N
O

G

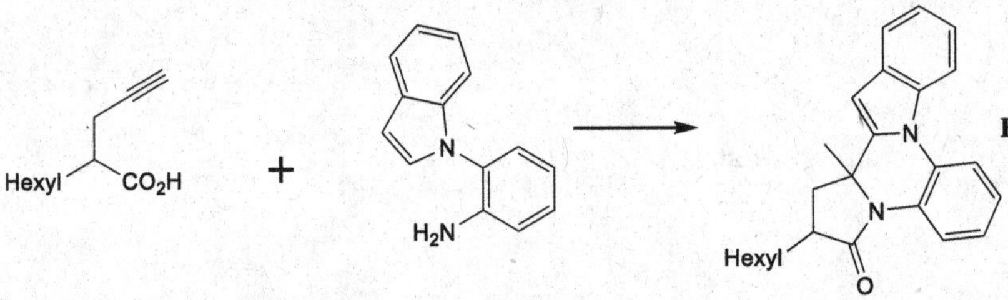
Hexyl
CO2H
+
N
H2N
N
N
Hexyl
O

H

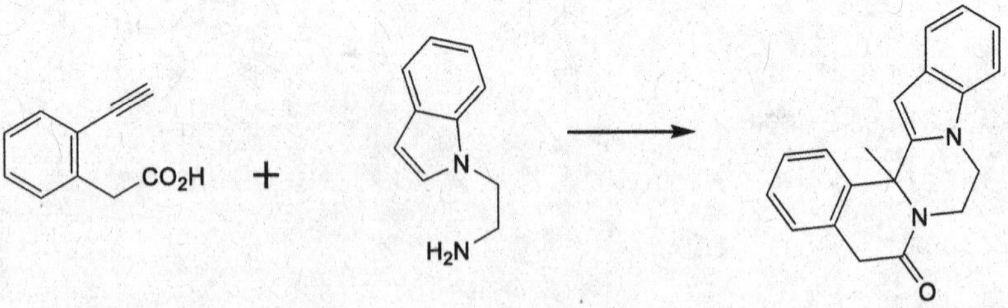
CO2H
+
N
H2N
N
N
O

I

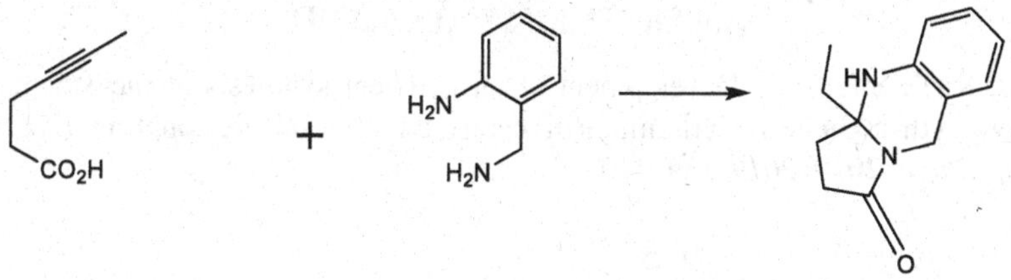
CO2H
H2N
H2N
HN
N
O

J

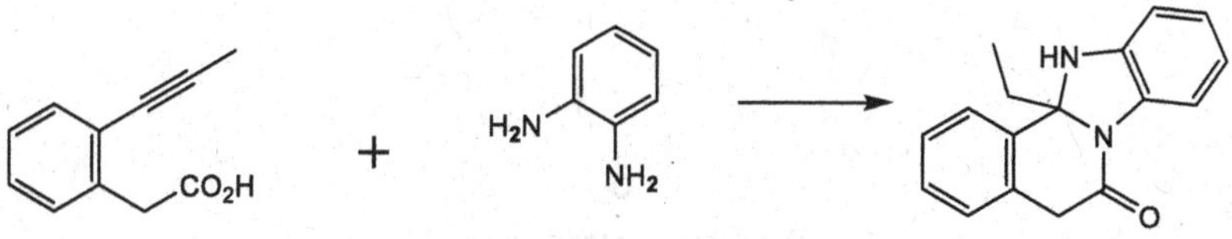
CO2H
H2N
NH2
HN
N
O

K

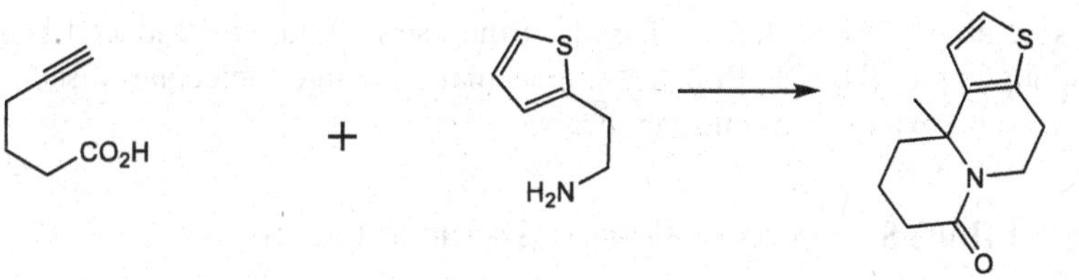
CO2H
S
H2N
S
N
O

L

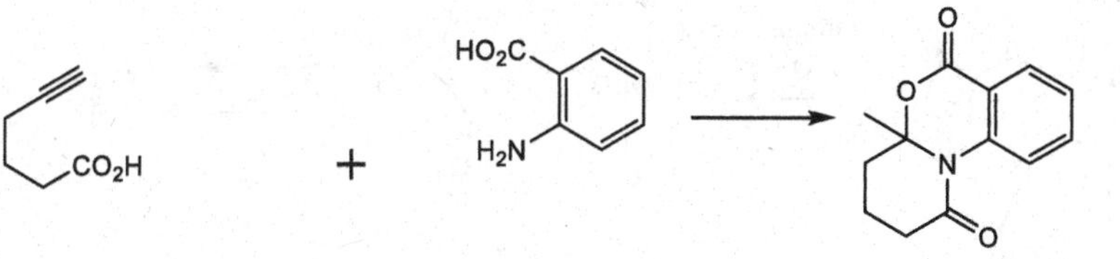
CO2H
HO2C
H2N
O
O
N
O

M

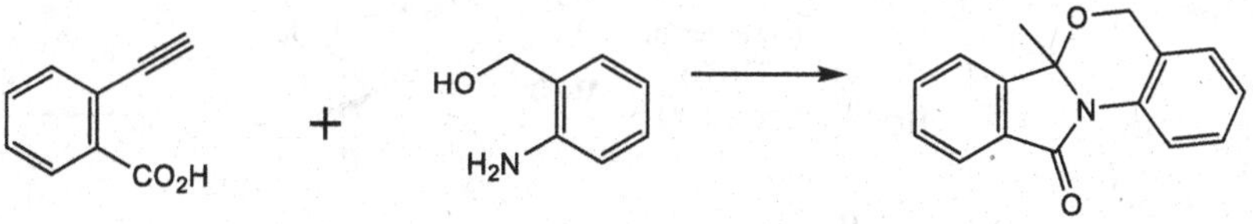
CO2H
HO
H2N
O
N
O

N

AN ERYTHRINA ALKALOID

Luu, R.-T.; Streuff, J. Development of an Efficient synthesis of rac-3-deme-thoxyerythratidione via a titanium(iii) catalysed imine-nitrile coupling. *Eur. J. Org. Chem.* **2019**, *2019*, 139–149

(+)-3-Demethoxyerythratidione   O

Synthesis of an erythrina alkaloid based on the use of Titanium Induced Imine-Nitrile Coupling (**TiIINC**). Both intramolecular and intermolecular disconnect approaches were explored, as outlined below.

PROJECTED TiIINC SEQUENCES OF POSSIBLE SYNTHETIC UTILITY

INTRAMOLECULAR TIINC APPROACH

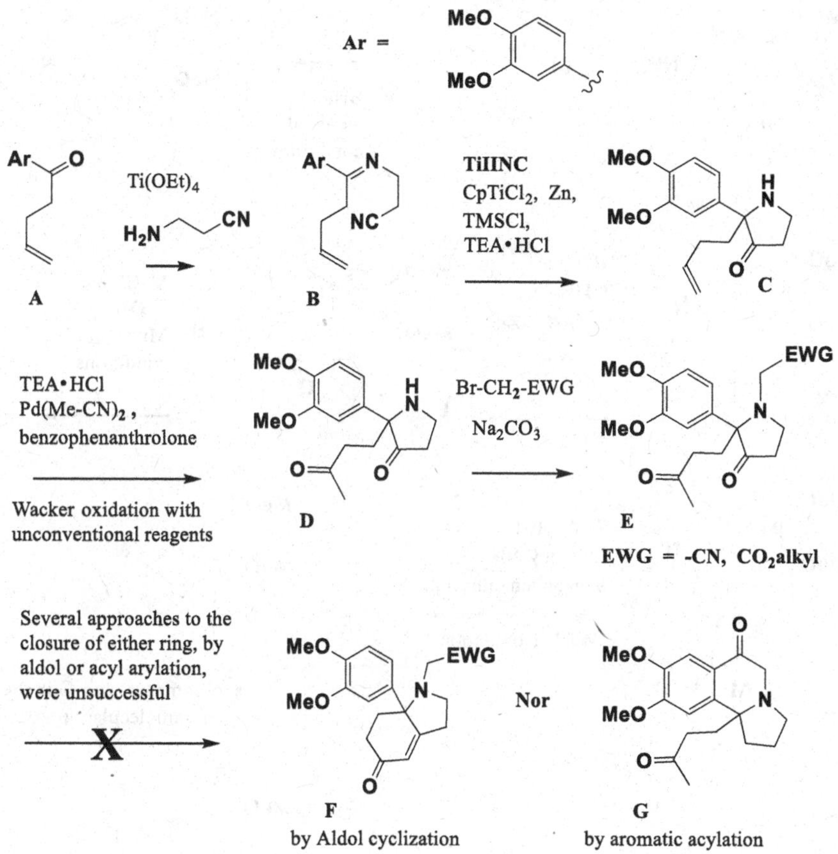
Ar =
MeO
MeO
Ar
O
Ti(OEt)4
H2N
CN
A
Ar
N
NC
B
TiINC
CpTiCl2, Zn,
TMSCl,
TEA•HCl
MeO
MeO
H
N
O
C
TEA•HCl
Pd(Me-CN)2 ,
benzophenanthrolone
Wacker oxidation with
unconventional reagents
MeO
MeO
H
N
O
O
D
Br-CH2-EWG
Na2CO3
EWG
MeO
MeO
N
O
O
E
EWG = -CN, CO2alkyl
Several approaches to the
closure of either ring, by
aldol or acyl arylation,
were unsuccessful
X
MeO
MeO
EWG
N
O
F
by Aldol cyclization
Nor
O
MeO
MeO
N
O
G
by aromatic acylation

The synthesis was realized using an intermolecular variation, shown below.

LNTERMOLECULER TIINC APPROACH

H   I   Bishler-Napieralski conditions   J

K   NC⁀OH **TiIINC** CpTiCl₂, Zn, TMSCl, TEA•HCl   L   MsCl, pyr **OR** Mitsunobu conditions

M   TEA•HCl Pd(Me-CN)₂, benzophenanthrolone   Wacker oxidation   N (compare with **G** in the intramolecular approach)

N   Aldol cyclization   O   (+)-3-demethoxy erythrotidione

ELEMINE

β-Elemene

Benito-Iglesias, D.; Herrero Teijon, P.; Rubio Gonzalez, R.; Fernandez-Mateos, A.Synthesis of trans- β -elemine. *Eur. J. Org. Chem.* **2018**, *2018*, 4926–4932

"...extensive and potent antineoplastic activity against a wide variety of tumors, including brain, breast, liver, lung and prostate as well as other tumors resistant to traditional drugs," Reviewed by M. Adio *Tet.* **2009**, *65*, 5145

SYNTHETIC METHOD

Limonene   A   J

advanced intermediate   β-Elemene

TiCRC = Titanium^{III} catalysed reductive cyclization

Trans- β -elemine is synthesized by two parallel sequences. Both utilize limonene as starting material for the preparation of the keto aldehyde **A**. **A** is then converted by two different sequences of the Corey–Chakowski epoxidation, and Homer–Wittig olefination to the common advanced intermediate **J**. A late stage funtionalization of **J**, using an elegant Ti^{III}-catalysed reductive cyclization (**TiCRC**) delivers the tetra-substituted cyclohexane core **K** of elemine, with the quaternary carbon and the other two chiral centers in the correct configuration. Oxidation of the primary alcohol, and olefination, completes the synthesis.

These two sequences are more than the same reactions done in a different order. Such variations are explored in search of higher yields, and alternate avenues towards derivitization.

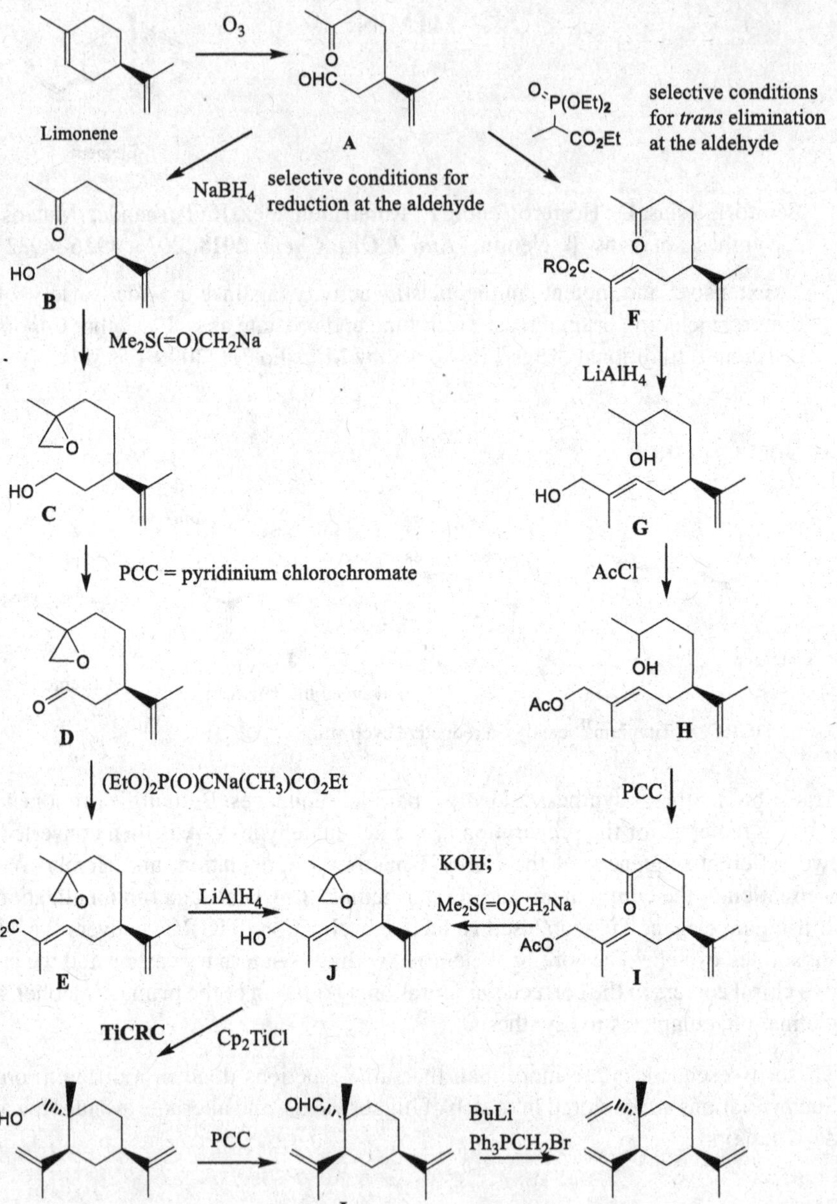
O3
OHC
O
Limonene
A
O=P(OEt)2
CO2Et
selective conditions
for trans elimination
at the aldehyde
NaBH4
selective conditions for
reduction at the aldehyde
HO
B
RO2C
F
Me2S(=O)CH2Na
LiAlH4
C
OH
G
PCC = pyridinium chlorochromate
AcCl
D
AcO
H
(EtO)2P(O)CNa(CH3)CO2Et
PCC
KOH;
Me2S(=O)CH2Na
LiAlH4
E
J
I
TiCRC
Cp2TiCl
OHC
PCC
BuLi
Ph3PCH2Br
K tetrasubstituted
cyclohexane core
L
β-Elemene

3-(R)-HYDROXYGLUTAMIC ACID

Piotrowska, D.G.; Glowaka, I.E.; Wrobleski, A.E.; Lubowiecka, L. Synthesis of nonracemic hydroxyglutamic acids. *Beilstein. J Org. Chem.* **2019**, *15*, 236–255

K

2(S)-amino-3(R)-hydroxyglutaric acid = 3(R)-hydroxyglutamic acid

Three distinct synthetic approaches are used to establish the oxygen chirality at the β carbon. These use Garner's aldehyde **A** and serine **H** as starting materials in which the chirality of the nitrogen is prepositioned.

In the first, the core of Garner's aldehyde acts as a chiral auxiliary in directing its own stereospecific oxa-Diels–Alder addition to Danashevsky's diene.

The second synthesis uses the chirality of Garner's aldehyde to direct the enantioselectivity of a Grignard attack on its own carbonyl group.

Serene itself is the starting material in the third synthesis. It is initially protected at the carboxylic acid and amine, then oxidized to the aldehyde. The Fmoc protected chiral amine then guides a classic Reformatsky addition, producing both the chiral protected alcohol and the second carboxylate group.

Standard conditions are used to hydrolyze the esters and install or remove the protecting groups.

THE THREE SYNTHETIC APPROACHES

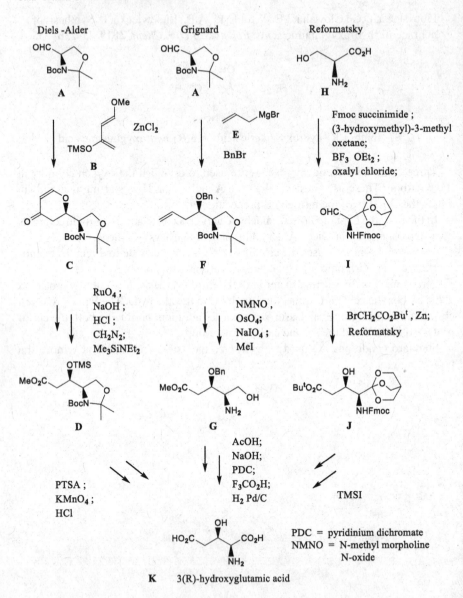

Diels -Alder

Grignard

Reformatsky

K   3(R)-hydroxyglutamic acid

PDC = pyridinium dichromate
NMNO = N-methyl morpholine
          N-oxide

(R) AND (S) 3-(N-PROPYL)-BUTYROLACTONE

Ciceri, S.; Grisenti, P.; Reza Elahi, S.; Ferraboschi, P. A new chemoenzymatic synthesis of the chiral key intermediate of the antiepileptic brivaracetam. *Molecules* **2018**, *23*, (9), 2206

Brevaracetam

Chiral Key Intermediate (R)-J

(R)-3-(n-propyl)-butyrolactone

The production of Brevaracetam was based on the use of **(R)-J** at an early stage of the synthesis. It was necessary to obtain this key intermediate in high yield and purity.

KNOWN THREE-STEP PREPARATION OF (S)-J

The ingenious and concise synthesis below was known to deliver the undesired enantiomer **(S)-J** from (R)-(-) epichlorohydrin.

The low yield reported here prompted the development of the following chemoenzymatic route for the preparation of the desired (R) enantiomer.

CHEMOENZYMATIC ROUTE TO THE (R)-J ENANTIOMER

A better yielding multistep procedure was designed, using a phenyl ring as a convertible surrogate for the carboxylic acid functionality. Racemic alcohol rac-E was prepared with a Perkin reaction through **C** and **D**. Enzymatic resolution of rac-E with *pseudomonas florescens* lipase (PFL) established the correct chiral center in **(R)-E**. The opposite acetate **(S)-E** was recycled.

(R)-E was acetylated and then oxidized with ruthenium tetroxide to **(R)-G**

(R)-G could then be attached to a chiral auxiliary ((R)-(+)-phenylethylamine), producing diastereomer **(R)-I**, which was then used for chiral analysis

The final steps involved hydrolysis of the **(R)-G** to **(R)-H**, then lactonization to **(R)-J**

C R = H } EtOH
D R = Et } H₂SO₄

rac-E

(R)- E
99% ee, 31% yield

(S)- E

(R) G X = Ac } K₂CO₃
(R) H X = H } MeOH

Used for spectral configuration analysis and large scale purification

(R) I

(R) J   (R)-3-(n-propyl)-butyrolactone (99% ee, 56% from (R)-E)

INDOLE – QUINOLINE HYBRIDS

Li, W.; Shuai, W.; Sun, H.; Xu, F. ...Xu, S. Design, synthesis and biological evaluation of quinoline-indole derivatives as anti-tubulin agents targeting the colchicine binding site. *Eur. J Med. Chem.* **2019**, *163*, 428–442

THE TARGET HYBRIDS

The Target Hybrids

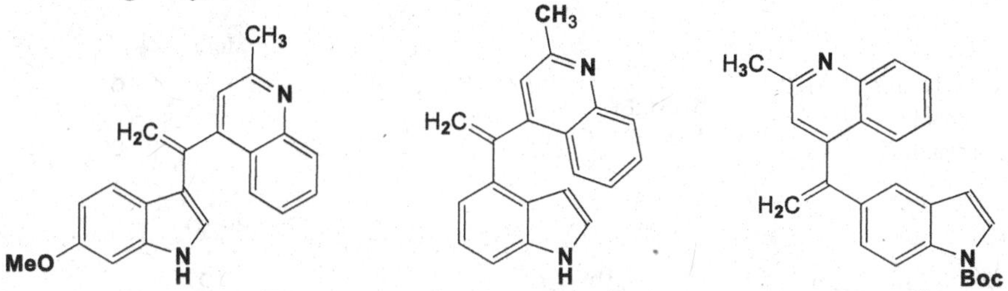

A   B   C

SYNTHESIS OUTLINE

The acetylation of indole at C-3, C-4, and C-5 is carried out by three different methods. A palladium catalyzed coupling between 4-chloro-2-methyl quinoline and the hydrazones (**K, L, M**) of these acetylated indoles delivers the hybrids **A, B,** and **C**.

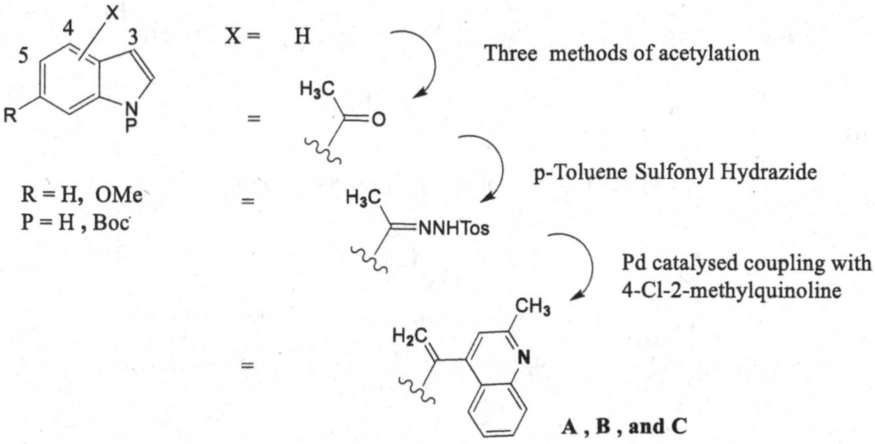

R = H, OMe
P = H , Boc

X = H

= Three methods of acetylation

= p-Toluene Sulfonyl Hydrazide

= Pd catalysed coupling with 4-Cl-2-methylquinoline

A , B , and C

Hybrids Were Derived at the Three Indole Positions from Starting Materials D, F, and H

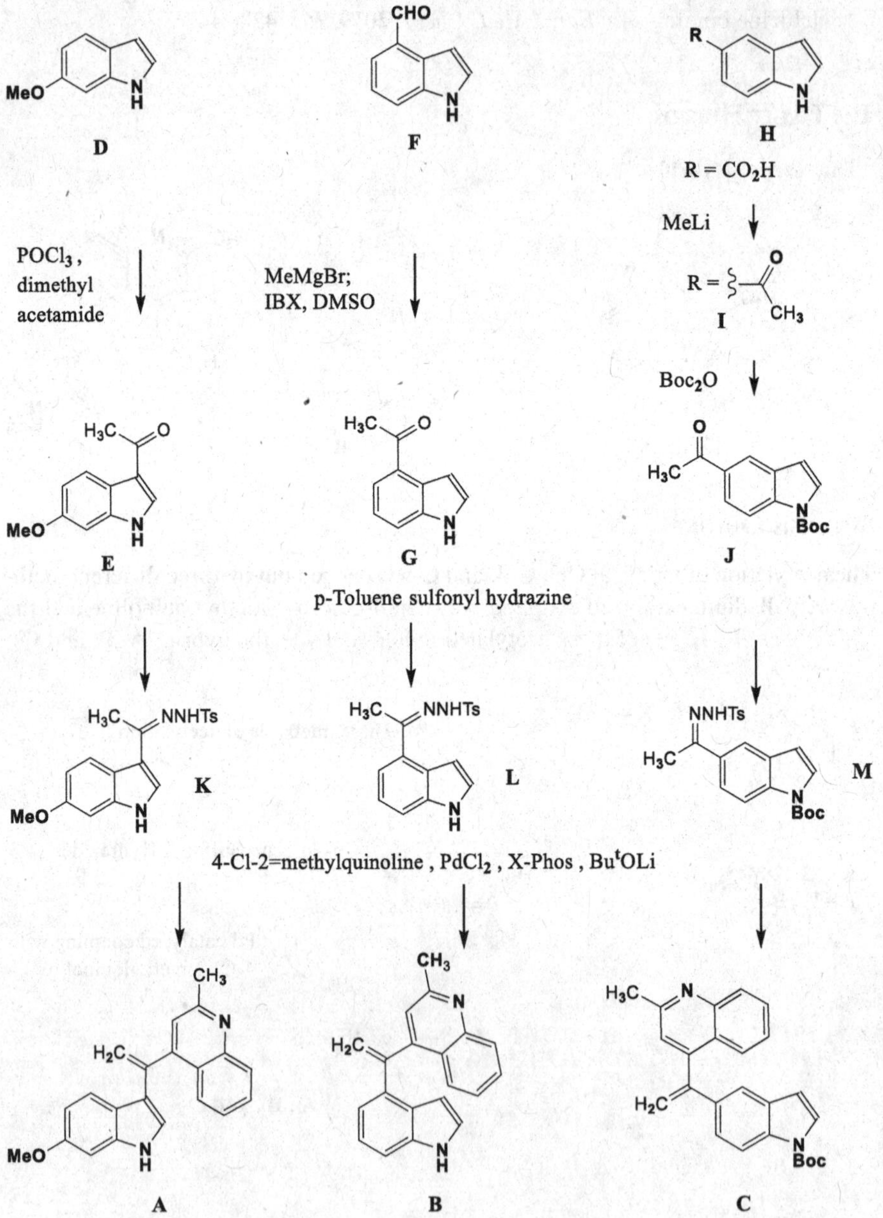

D

F

H

R = CO$_2$H

MeLi

R = $\overset{O}{\overset{\|}{C}}CH_3$

I

POCl$_3$,
dimethyl
acetamide

MeMgBr;
IBX, DMSO

Boc$_2$O

E

G

J

p-Toluene sulfonyl hydrazine

K

L

M

4-Cl-2=methylquinoline , PdCl$_2$, X-Phos , Bu^{t}OLi

A

B

C

ISOQUINOLINE ALKALOID

Melzer, B. C.; Felber, J. G.; Bracher, F. Aminomethylation/hydrogenolysis as an alternative to direct methylation of metalated isoquinolines – a novel synthesis of the alkaloid 7-hydroxy-6-methoxy-1-methylisoquinoline. *Beilstein J. Org. Chem.* **2018**, *14*, 130–134

7-hydroxy-6-methoxy-1-methylisoquinoline

Four syntheses (I, II, III, and **IV**) of this simple compound are outlined. Only the last (**IV**) is satisfactory with respect to ease of preparation, purification efficiency, and yield.

I

Aromatization here is quite inefficient

II

This tandem Bishler–Napieralsky reaction and hydrogenative debenzylation unexpectedly and fortuitously aromatize the nitrogen-containing ring, forming **A**, unfortunately in low yield. The methyl group was implicated in the inefficiency of the aromatization.

III

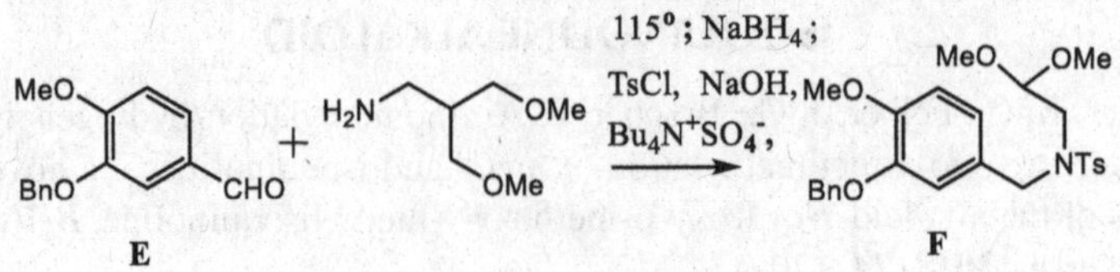

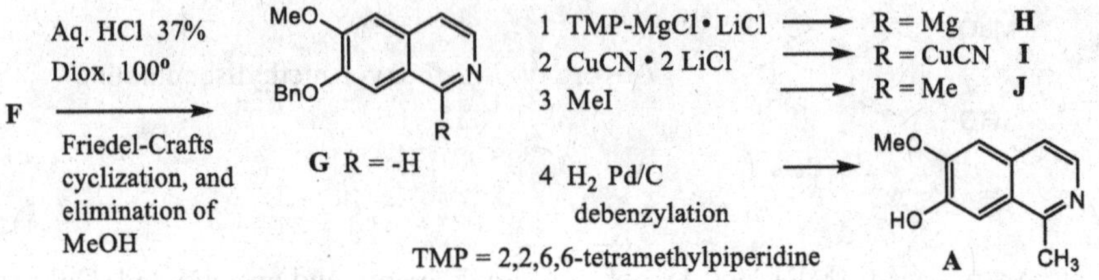

$115°$; $NaBH_4$;
$TsCl$, $NaOH$, MeO,
$Bu_4N^+SO_4^-$;

E + (amine reagent) → **F**

Aq. HCl 37%
Diox. $100°$

F →

Friedel-Crafts
cyclization, and
elimination of
MeOH

G R = -H

1 TMP-MgCl • LiCl ⟶ R = Mg   **H**
2 CuCN • 2 LiCl ⟶ R = CuCN   **I**
3 MeI ⟶ R = Me   **J**

4 H_2 Pd/C
debenzylation

TMP = 2,2,6,6-tetramethylpiperidine

→ **A**

Debenzylation of **J** to **A** was inefficient, as it was in example **II** (**B** to **A**). This was thought to be due to an electronic effect from the methyl group adjacent to the nitrogen. A "masked methyl group" was implemented in the approach shown below in **IV**.

IV

The dimethylaminomethyl group was installed as substituent R with Eschinmoser's salt and cyanocuprate **I**, forming **K**. Methylation with methyl iodide yielded the trimethyl ammonium iodide salt **L**, but **L** did not debenzylate well. It was converted into the chloride **M**, which debenzylated cleanly during the "unmasking" procedure, yielding the product **A** in acceptable yield and purity.

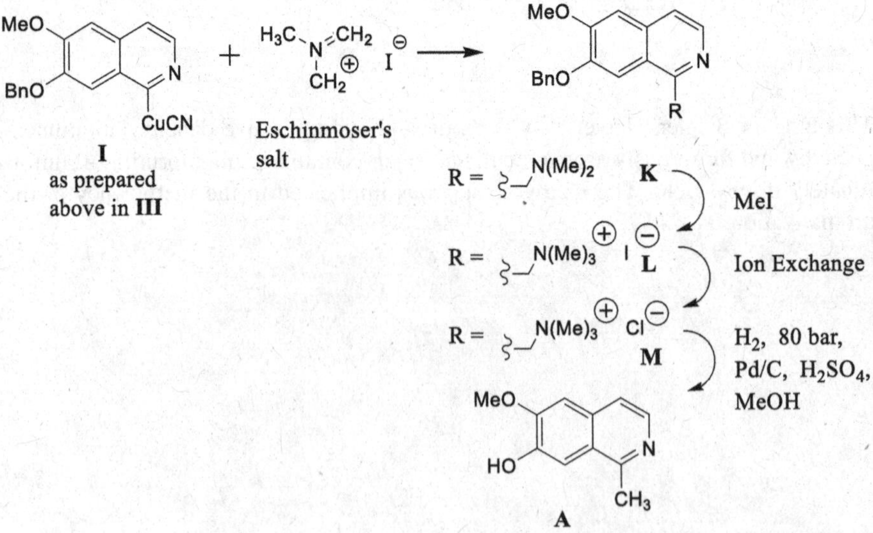

I
as prepared
above in **III**

Eschinmoser's
salt

R = ∿N(Me)$_2$ **K**

MeI

R = ∿N(Me)$_3^{\oplus}$ I$^{\ominus}$ **L**

Ion Exchange

R = ∿N(Me)$_3^{\oplus}$ Cl$^{\ominus}$ **M**

H_2, 80 bar,
Pd/C, H_2SO_4,
MeOH

A

LILY-OF-THE-VALLEY ODORANT

Jordi, S.; Kraft, P. Crossing the boundries between marine and muguet: Discovery of unusual lily-of-the-valley odorants devoid of aldehyde functions. *Helvetica Chim. ACTA* **2018**, *101* (6), e1800048

"Muguet" is a lily-of-the-valley perfume. Synthetic fragrances with aldehyde functionalities lose their aroma within the usage lifetime of the product. These syntheses were realized to overcome that particular shortcoming in muguet.

Note the use of both classical and modern cyclopropanation reagents in the preparation of intermediates E and K.

SYNTHESIS OF A

Target Structures

A n = 0, R = CH$_3$
B n = 1, R = H

Synthesis of A

C   X = CHO

EtPPh$_3$

D   X =

CH$_2$I$_2$, Et$_2$Zn Simmons-Smith reagent

E   X =

F

BBr$_3$, Me$_2$S ; Cl ⟍⟍ Cl

A

[RuO$_4$]

The diasteriomers are inseparable ; benzylic protons appear at δ(H) 1.92 and 2.11

SYNTHESIS OF B

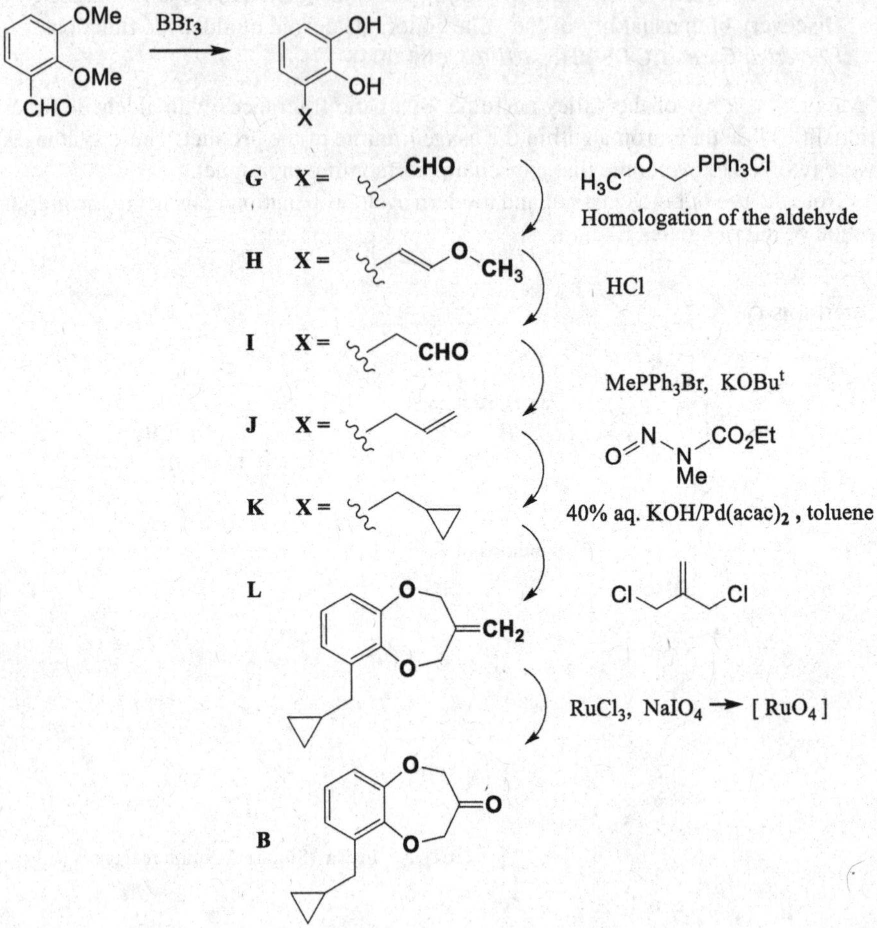

G   X =   CHO

H₃C–O–CH₂–PPh₃Cl

Homologation of the aldehyde

H   X =   O–CH₃

HCl

I   X =   CHO

MePPh₃Br, KOBuᵗ

J   X =

O=N–N(Me)–CO₂Et

K   X =

40% aq. KOH/Pd(acac)₂ , toluene

L

Cl   Cl

RuCl₃, NaIO₄ → [RuO₄]

B

ENZYME RECEPTOR LIGAND

Doebelin, C.; Patouret, .; Garcia-Ordonez, R. D.; Chang, M. R.; Dharmarajan, V.; Novick, S.; Ciesla, A.; Campbell, S.; Solt, L.A.; Griffin, P.R.; Kamenecka, T. M. Identification of potent RORβ modulators: Scaffold variation. *Bioorg. Med. Chem. Lett.* **2018**, *28*, 3210–3215

The **R**etinoic **A**cid **R**eceptor is an enzyme that is activated by a known variety of naturally occurring and synthetic hormones. That receptor binds compound **A**, whose structure belongs to that within the **Generalized Scaffold** structure shown below. Compound **A** is a "hit": a registered candidate with therapeutic activity.

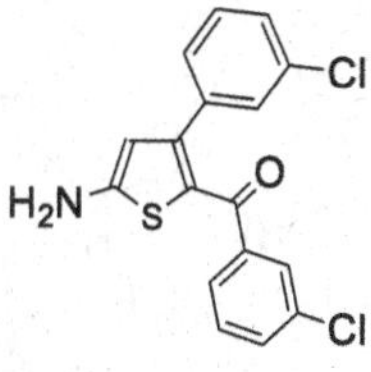

A

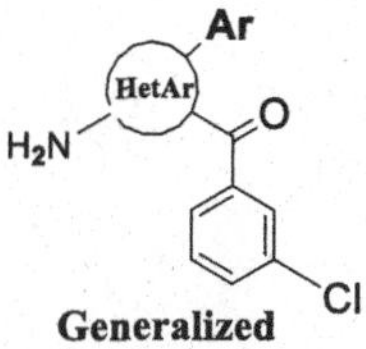

Generalized Scaffold Types

Compound **A** affects the metabolic cycles that involve the Retinoic Acid Receptor. In order to mediate, moderate, or inhibit these metabolic cycles, a synthesis of the **Generalized Scaffold** was needed. The synthesis of compound **A** shown below was taken as a starting point for such syntheses.

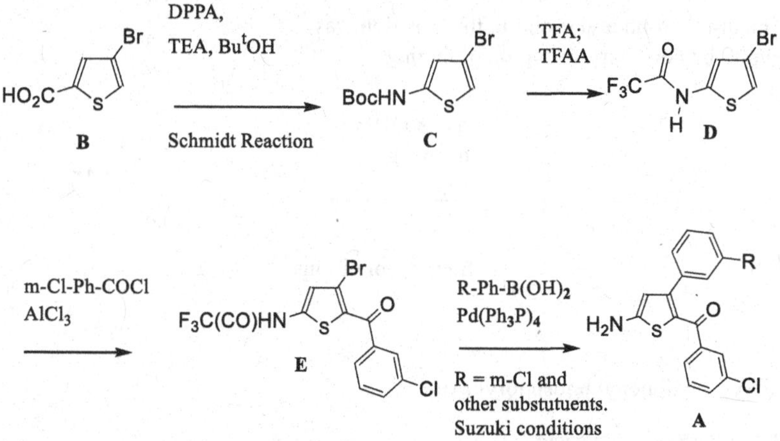

Based on the synthesis of **A**, several analogs were prepared using this stylized outline

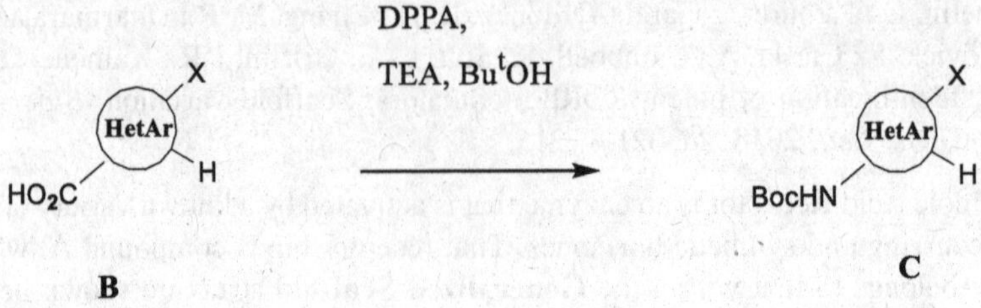

DPPA,

TEA, Bu^{t}OH

HetAr

X

H

HO$_2$C

B

BocHN

HetAr

X

H

C

X = Br or OTf

TFA; TFAA

C ⟶

F$_3$C

O

N
H

HetAr

X

H

D

m-Cl-Ph-COCl

AlCl$_3$

D ⟶

Friedel Crafts
acylation

F$_3$C

O

N
H

HetAr

X

O

Cl

E

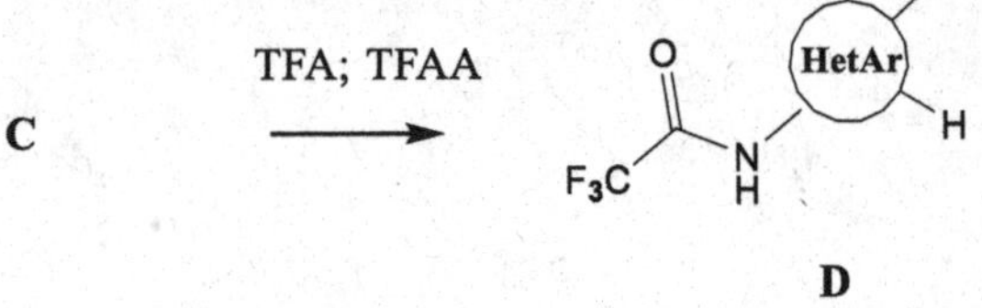

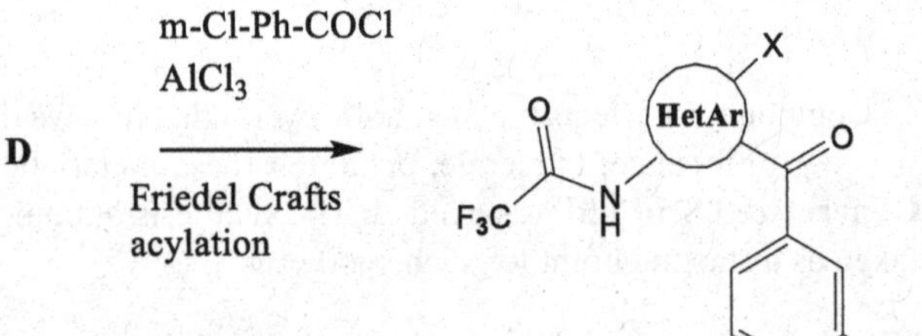

The meta-chlorophenyl group in this position was
known to be necessary for optimal binding.

Ar-B(OH)$_2$

Pd(Ph$_3$P)$_4$

E ⟶

Suzuki conditions

H$_2$N

HetAr

Ar

O

Cl

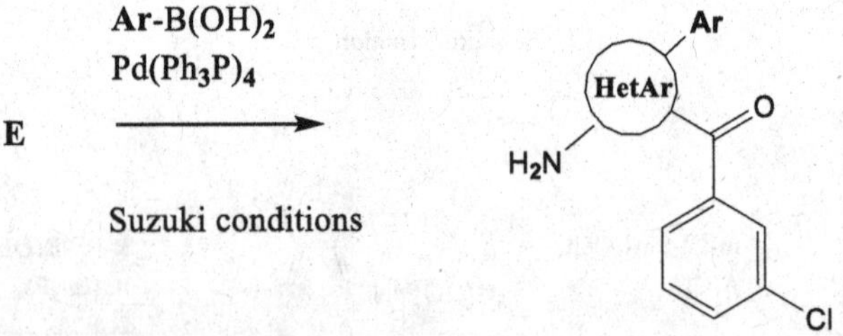

DPPA   = diphenyl phosphoryl azide

TFA = trifluori acetic acid

TFAA = trifluoro acetic anhydride

The candidate analogs are "Retinoic Acid Receptor Related Orphan Binders," referred to informally as "ROR (β & γ)." Variations on the **Generalized Scaffold** were prepared with combinations of the following functionalities in the **HetAr** and **Ar** positions.

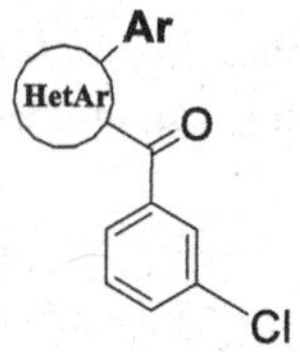

Generalized Scaffold

Ar =

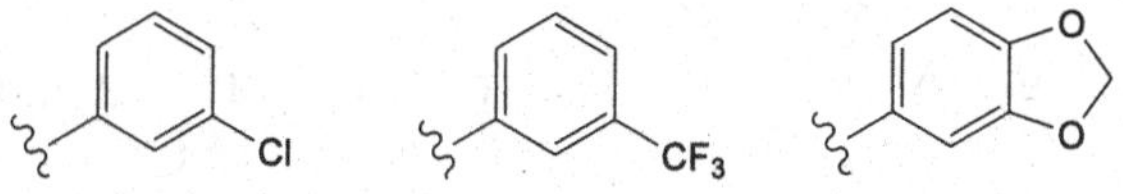

A

HetAr =   Twenty heterocycles, including :

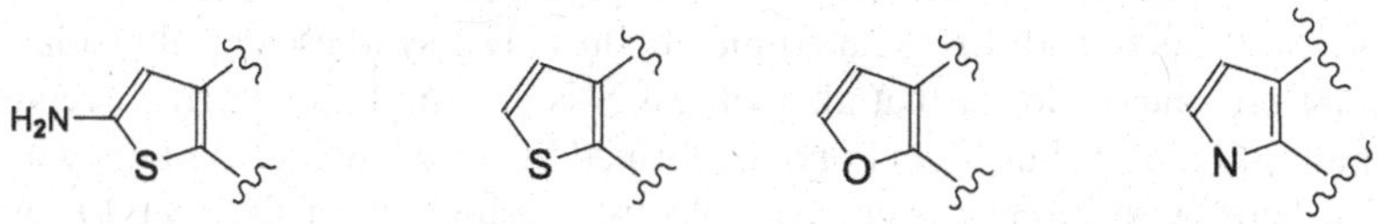

A

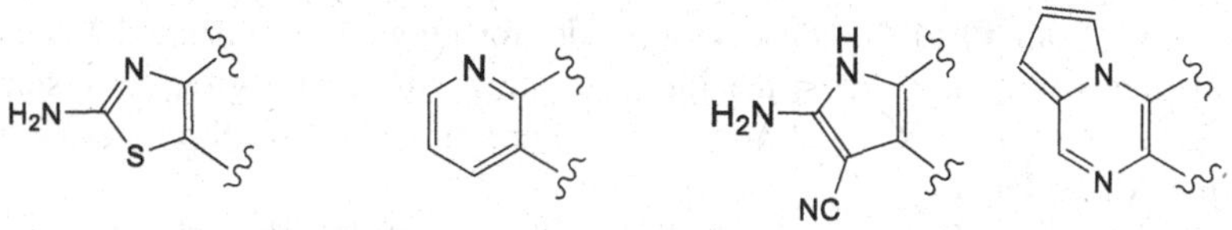

This report documents the preparation, binding constants, and relevant metabolic effects of this library of analogs. Analysis indicated the influence of the **HetAr** and **Ar** moieties on selected pharmacokinetic features.

ALTERSOLANOLS A AND N

Mechsner, B.; Bose, D.; Hogenkamp, F., Ledermann, N.; ... Pietruska, J. Enantioselective total synthesis of altersolanol A and N *Bioorg. and Med. Chem.* **2019**, *27*, 2991–2997

Mechsner, B.; Henssen, B.; Pietruszka, J. First enantioselective total synthesis of altersolanol A. *Org. Biomol. Chem.* **2018**, *16*, 7674–7681

	R^1	R^2	R^3	R^4	R^5	
Altersolanol A	H	OH	OH	H	OH	
B	H	H	OH	H	H	
C	H	OH	OH	H	H	hydrolysis
M	OH	H	H	OAc	OH	
N	H	OH	OAc	H	OH	

The Altersolanols

A practical synthesis of the altersolanols was needed to assay the biomedical activity of these compounds and their related metabolites. The syntheses of altersolanol A and N are presented. Notice that altersolanol N is a naturally occurring acetyl version of altersolanol A, but N could not be formed by acetylation of A. The two were prepared here by different routes. The selective positioning of the acetyl group in altersolanol N, at the C-2 hydroxyl (R^3), was facilitated by an intramolecular acetyl migration (from **P**) to the adjacent epoxide, forming (the presumed) acetonide intermediates **Q** and **R**. From the intermediate **R**, the acetyl group is restored to its original position at R^3 by elimination of a proton and the TBS protecting group, forming Altersolenol N.

Authentic altersolenol A is formed by hydrolysis of this altersolanol N, making these two sequences essentially two distinct syntheses of altersolanol A.

INITIAL ALTERSOLANOL A SYNTHESIS

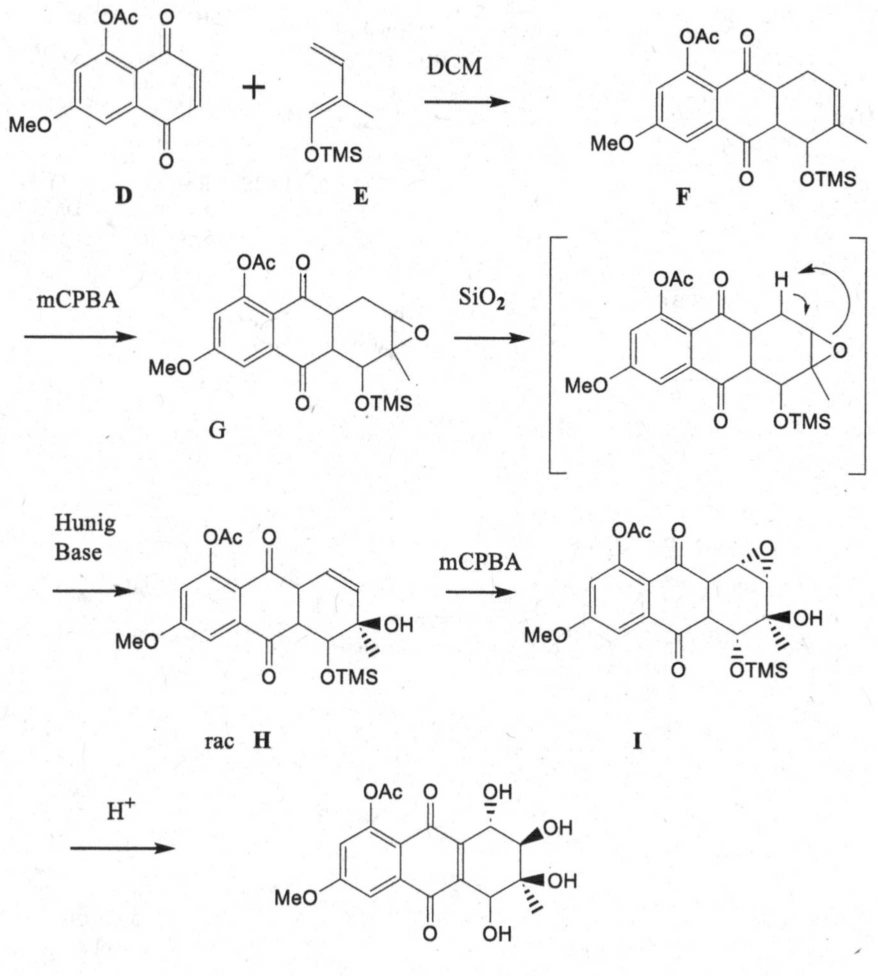

Altersolanol A

The selective direct acetylation of the C-2 hydroxyl group of Altersolanol A, to form Altersolanol N, was not realized. An alternate approach to the synthesis of the N form was developed by prepositioning the acetyl group at C-2, as shown below.

ALTERSOLANOL N SYNTHESIS, AND HYDROLYSIS TO ALTERSOLANOL A

This final conversion of Altersolanol N into the A form confirms the structural relationship between the two.

TENERAIC ACID

"A naturally occurring imino acid."

Amino, Y., Nishi, S., Izawa, K. Stereo-Selective preparation of teneraic acid, trans-(2s,6s)-piperidine-2,6-dicarboxyllic acid, via anodic oxidation and cobalt-catalyzed carbonylation. *Chem. & Pharm. Bull.* **2017**, *65*, 854–861

TENERAIC ACID AND ITS DIASTEREOMERS

A — Trans 2S, 6S
Target
"(2S,6S) Teneraic Acid"

B — Trans 2R, 6R

C — Cis-Meso 2S, 6R

THE ORIGINAL ONE-STEP SYNTHESIS OF TENERAIC ACID

Formylation of the N-benzoyl unsaturated amino acid **D** yielded two diastereomers of Teneraic acid. Previous work suggested the following! carbonylation mechanism for this one-pot reaction.

Although this reaction was useful in that it formed the desired carbon framework in a single step, it did not proceed stereospecifically. The intermediacy of N-acyliminium ion **F** was suspected of involvement in the isomerization.

Revised Two-Stage Approach to Teneraic Acid

In view of the possible influence of the N-acyliminiun ion F on the isomerization at **G**, a new two-step synthesis was planned with a different starting material. The transitional iminium ion **K** would be generated from the ester N-benzoyl methyl pipecolate **J**, and trapped as methyl ether **L**, without immediately proceeding to the carbonylation, as done before in the one-step synthesis with intermediate **F**.

Methyl ester **L** was then carbonylated under cobalt catalysis. The N-acyliminium ion **K** (identical to the ion from which the methyl ether **L** was formed) is presumed to be an intermediate in the carbonylation to follow according to the mechanism **L** to **M** to **N**. The products of this carbonylation were both the N-benzoyl acid ester **O** and diester **P**, due to the absence of enough methanol.

Treatment of this **O/P** mixture under severe conditions (refluxing concentrated HCl) deprotected the nitrogen and hydrolyzed the esters, yielding an acceptably pure teneraic acid **A**.

PACHASTRISSAMINE

"A cytotoxic anhydrophytosphingosine isolated from a marine sponge."

Two retrosynthetic approaches to pachastrissamine are outlined below

DIETHYL TARTARATE APPROACH

Fujiwara, T.; Liu, B.; Niu, W.; Hashimoto, K.; Nambu, H. Practical synthesis of pachastrissamine (jaspin b), 2-epi-pachastrissamine, and the 3-epi-pyrrolidine analog. *Chem. Pharm. Bull.* **2016** *64*, 179–188

| J pachastrissamine | F | D Triol | DET diethyl tartarate |

Cyclization of **D** to the tetrahydrofuran core is realized by selective tosylation of the primary alcohol, with immediate displacement of the tosylate by the homoallylic alcohol, forming **F**.

GARNER'S ALDEHYDE APPROACH

Lee, H.-J.; Lim, C.; Hwang, S.; Prof. Jeong, B.S.; Prof. Kim, S. Silver-mediated exo-selective tandem desilylative bromination/oxycyclization of silyl-protected alkynes: Synthesis of 2-bromomethylene-tetrahydrofuran.

Chem. Asian J **2011**, *6*, 1943

| J pachastrissamine | M | K | Garner's Aldehyde |

Deprotection of the acetonide and TIPS groups in **K** allows the formation of the tetrahydrofuran core. Bromination of the acetylene group is followed by the addition of the primary alcohol to the acetylene, forming **M**.

DIETHYL TARTARATE EXPERIMENTAL

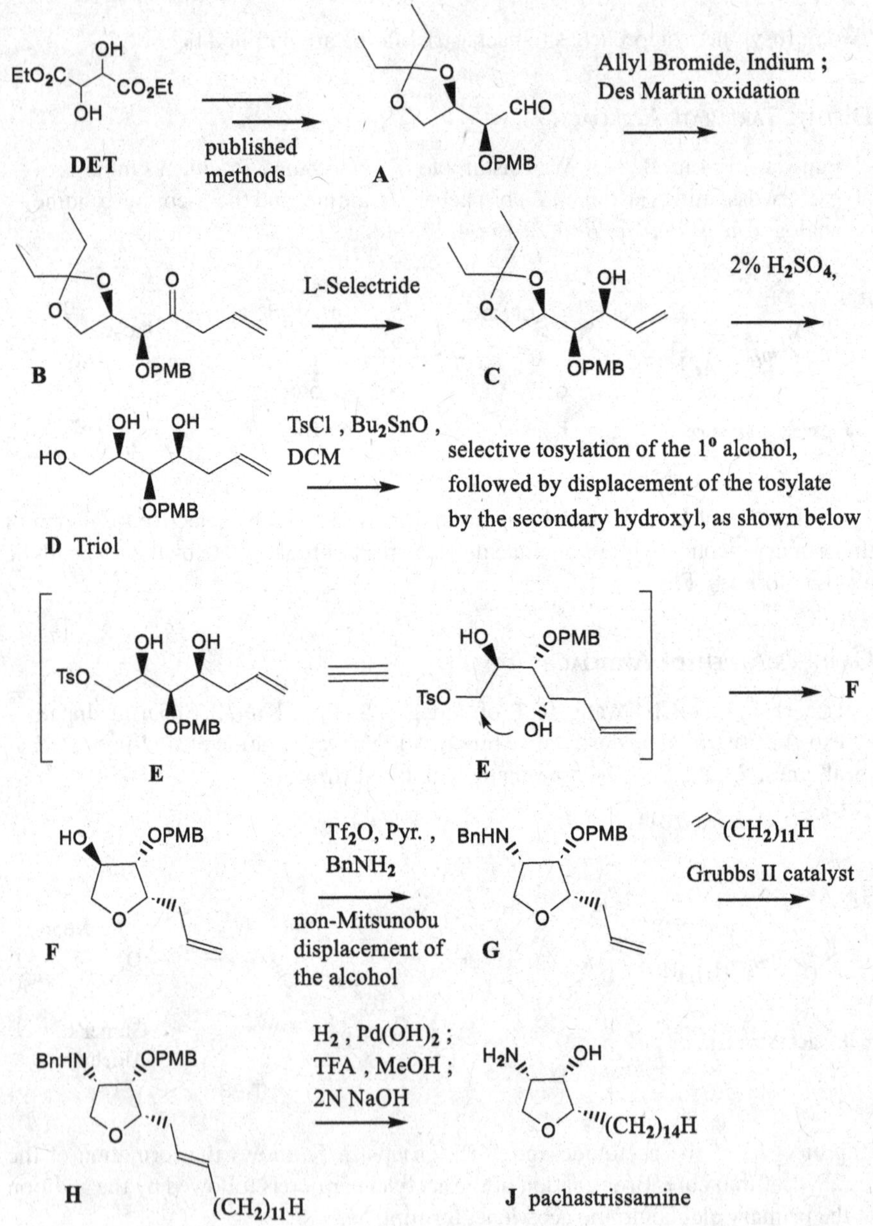

DET → published methods → A

Allyl Bromide, Indium ;
Des Martin oxidation →

B + L-Selectride → C

2% H$_2$SO$_4$, →

D Triol

TsCl , Bu$_2$SnO ,
DCM →

selective tosylation of the 1° alcohol,
followed by displacement of the tosylate
by the secondary hydroxyl, as shown below

E ≡ E → F

F

Tf$_2$O, Pyr. ,
BnNH$_2$
→
non-Mitsunobu
displacement of
the alcohol

G

$\diagup$(CH$_2$)$_{11}$H

Grubbs II catalyst →

H

H$_2$, Pd(OH)$_2$;
TFA , MeOH ;
2N NaOH
→

J pachastrissamine

GARNER'S ALDEHYDE EXPERIMENTAL

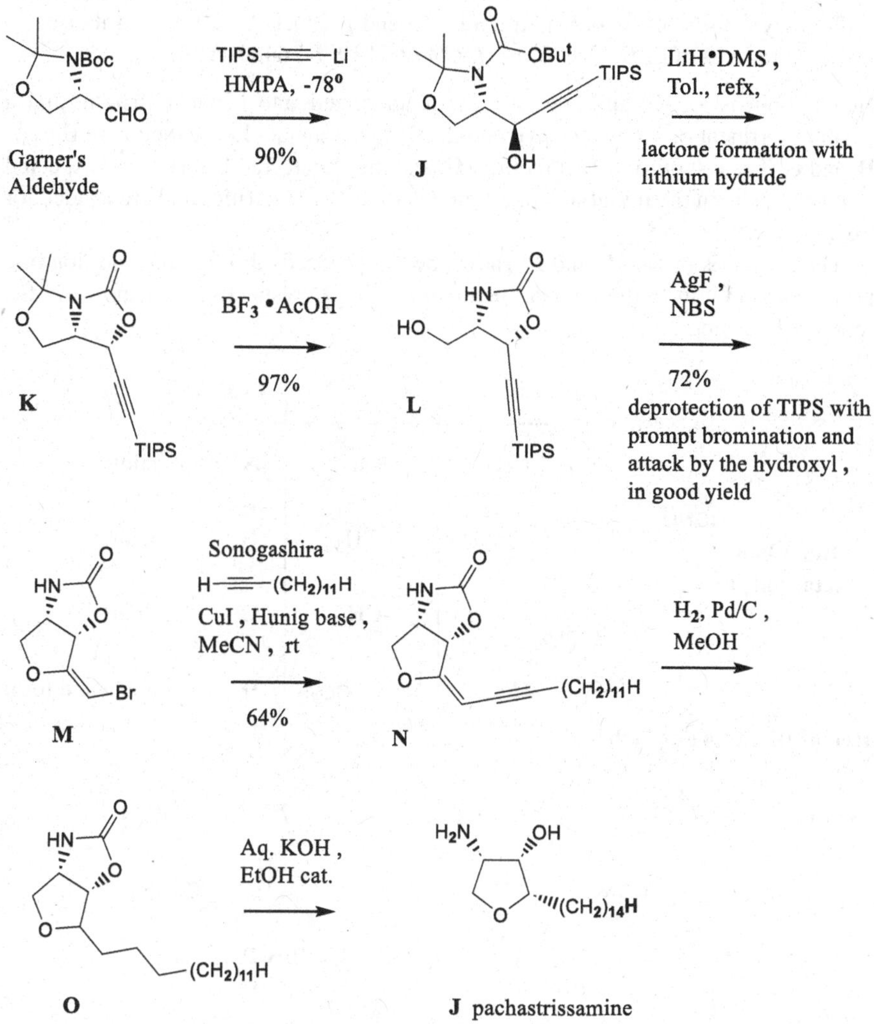
NBoc
CHO
Garner's Aldehyde
TIPS—≡—Li
HMPA, -78°
90%
O
OBu^t
N
TIPS
OH
J
LiH•DMS,
Tol., refx,
lactone formation with lithium hydride
K
TIPS
BF₃•AcOH
97%
HN
HO
L
AgF,
NBS
72%
deprotection of TIPS with prompt bromination and attack by the hydroxyl, in good yield
Br
M
Sonogashira
H—≡—(CH₂)₁₁H
CuI, Hunig base,
MeCN, rt
64%
(CH₂)₁₁H
N
H₂, Pd/C,
MeOH
(CH₂)₁₁H
O
Aq. KOH,
EtOH cat.
H₂N
OH
(CH₂)₁₄H
J pachastrissamine

BILIRUBEN METABOLITES

Schulze, **D.**; Traber, **J.**; Ritter, M.; Gorls, **H.**; Pohnert, G.; Westerhausen, M.

Total syntheses of the bilirubin oxidation end product Z-BOX C and its isomeric form Z-BOX D. *Org. Biomol. Chem.* **2019**, *17*, 6489–6496

Hemoglobin is slowly and continuously metabolized into heme, bilirubin, and a small assortment of secondary pyrrolyl-2(5H)-idene metabolites called here **Box A, B**, and **C**. The last of these metabolites, **C**, and the undetected **D** form, are suspected of involvement in the initiation of certain metabolic irregularities and are needed for toxicity testing.

The synthesis of **Box C** and **D**, shown below, proceeds through the addition of a propionic acid unit to the known precursors **K** and **O** respectively as shown in the retrosynthetic approach.

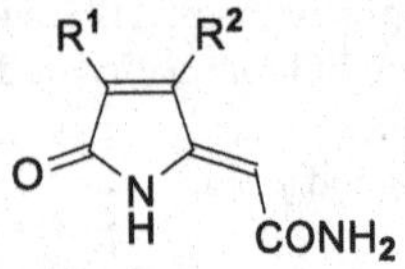

	R¹	R²	Box	
	$-CH_3$	Vinyl	**A**	major
	-Vinyl	$-CH_3$	**B**	major
	HO_2C~~~	$-CH_3$	**C**	minor
	$-CH_3$	~~~CO_2H	**D**	undetected form

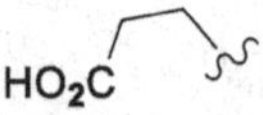

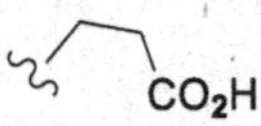

Biliruben Metabolites

Retrosnthetic Approach

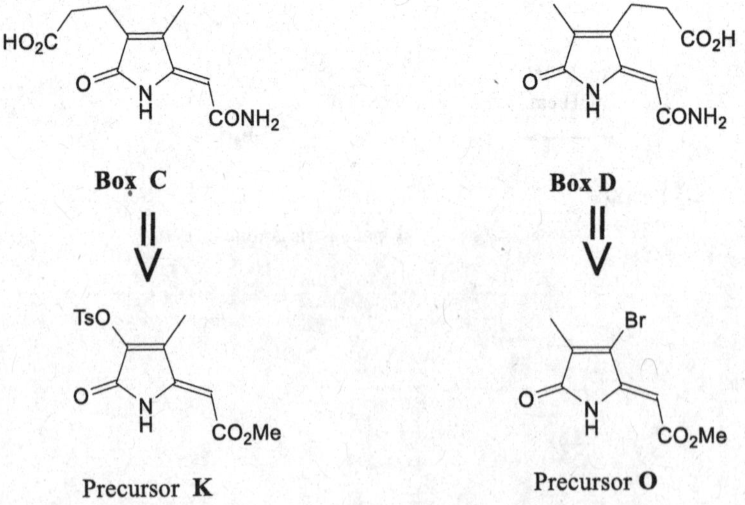

Box C

Box D

Precursor **K**

Precursor **O**

Conversion of the Precursors K and O into Box C and the Undetected Box D

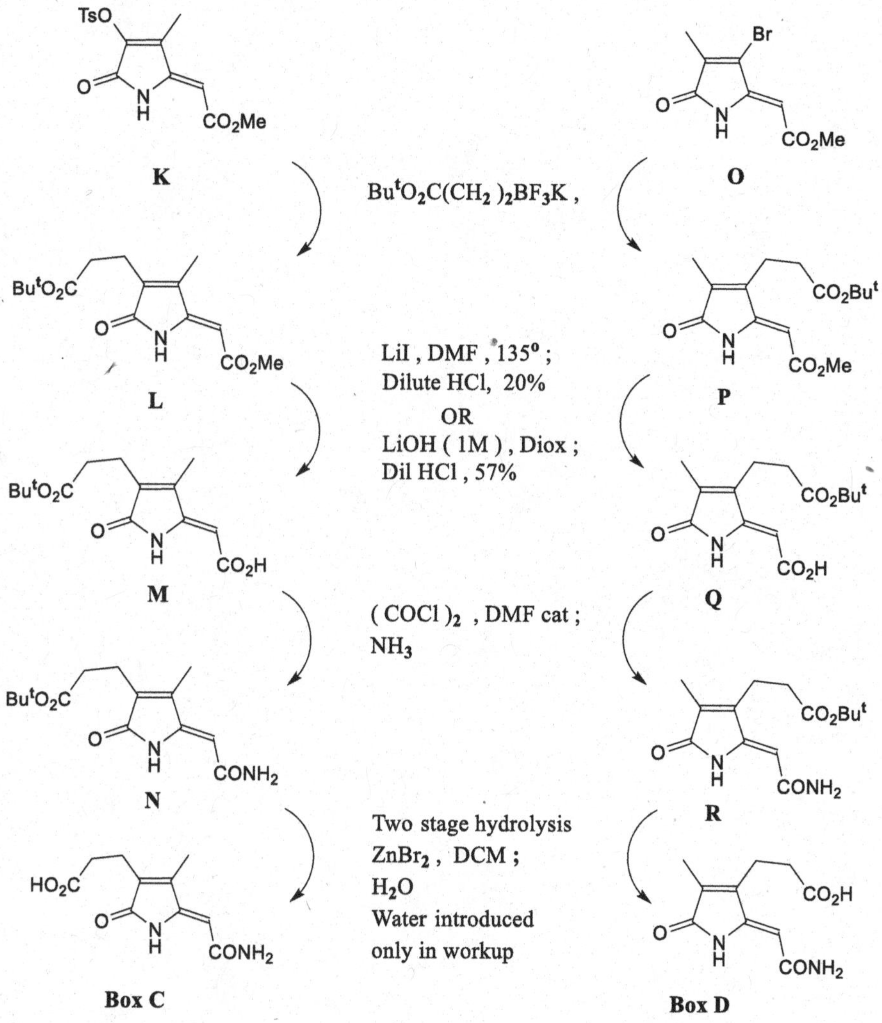
TsO
CO2Me
K
Br
CO2Me
O
ButO2C(CH2)2BF3K,
ButO2C
CO2Me
L
CO2But
CO2Me
P
LiI, DMF, 135°;
Dilute HCl, 20%
OR
LiOH (1M), Diox;
Dil HCl, 57%
ButO2C
CO2H
M
CO2But
CO2H
Q
(COCl)2, DMF cat;
NH3
ButO2C
CONH2
N
CO2But
CONH2
R
Two stage hydrolysis
ZnBr2, DCM;
H2O
Water introduced
only in workup
HO2C
CONH2
Box C
CO2H
CONH2
Box D

Index

NUMERICAL INDEX

ALPHEBETICAL INDEX

Printed in the United States
by Baker & Taylor Publisher Services